Airline Management Finance

Airline Management Finance: The Essentials is of significant benefit to airline industry practitioners seeking a focused, neatly contained and accessible resource that provides explicit financial information pertinent to their current or future role.

The book explains and demystifies an airline's financing and the financial reporting of its operations to airline staff and others. It seeks to explain the role of finance and the Finance Department in a non-technical way, so staff can appreciate the value of the department and its information resources, and see finance as an active contributor to the airline's operation. It concentrates on practical matters, explaining frequently used financial and accounting terms, how financial strategy works, the uses of various types of financial reporting, as well as what financial risk is and how it can be managed through the co-operation of finance and operating staff. Staff who understand the airline's finances and financial system are more likely to make decisions which align with the airline's strategy and objectives. They will also know how to use the financial information which is available. The book establishes a good foundation of financial knowledge for all staff.

This book is recommended reading for new employees in airline finance and related areas, as well as those starting to move up the supervisory ladder in an airline.

Victor Hughes is a Chartered Management Accountant who spent most of his career in asset-intensive industries and was Finance Director on the board of Cathay Pacific Airways Ltd. Victor designed and tutored an Aviation Financial Management course at Swinburne University of Technology, Melbourne, Australia.

Managing Aviation Operations
Series Editor: Peter J Bruce
Associate Editor: John M C King

The purpose of this series is to provide a comprehensive set of materials deal-ing with the key components of airline operations. To date, this innovative approach has not been evident among aviation topics and has certainly not been applied to the operational areas of airlines. While more recent works have begun, in brief, to consider the various characteristics of operational areas, the Managing Airline Operations for Aviation Professionals series will expand coverage with far greater breadth and depth of content.

Airlines are devoid of specific topic knowledge in ready-made, easy-to-read, creditable resources. Tapping into industry expertise to drive a range of key niche products will resource the industry in a way not yet seen in this domain. Therefore, the objective is to deliver a collection of specialized, internationally sourced and expertly written books to serve as readily accessible guides and ref-erences primarily for professionals within the industry. The focus of the series editors will be to ensure product quality, user readability and appeal, as well as transparent consistency across the range.

Airline Management Finance
The Essentials
Victor Hughes

For more information about this series, please visit: www.routledge.com/ Aviation-Fundamentals/book-series/MAO

Airline Management Finance

The Essentials

Victor Hughes

Routledge
Taylor & Francis Group

LONDON AND NEW YORK

First published 2020
by Routledge
2 Park Square, Milton Park, Abingdon, Oxon OX14 4RN

and by Routledge
52 Vanderbilt Avenue, New York, NY 10017

Routledge is an imprint of the Taylor & Francis Group, an informa business

British Library Cataloguing-in-Publication Data
A catalogue record for this book is available from the British Library

Library of Congress Cataloging-in-Publication Data
A catalog record has been requested for this book

ISBN: 978-1-138-61066-8 (hbk)
ISBN: 978-1-138-61069-9 (pbk)
ISBN: 978-0-429-46565-9 (ebk)

Typeset in Bembo
by Deanta Global Publishing Services, Chennai, India

Contents

Introduction

Aviation is an industry with a long history. It can be argued that aviation as an industry started with the commercial flights of dirigibles, which are balloons that can be steered. The aviation industry covers a wide range of activities, from carrying passengers and cargo to aerial photography and crop-spraying.

Airlines are the core part of the aviation industry and they provide an essential service to passengers and cargo shippers. Despite the important contribution airlines make to the national and the international economy, history shows that it is very difficult for an airline to meet passengers' and shippers' needs and consistently make a financial return. Operating an airline successfully requires a considerable investment of money, skills and people's time. Having an efficient operation is not just 'nice to have' for an airline, it is essential. There have been remarkable changes over the years in the way airlines operate and in the financial resources needed by airlines.

The success and survival of an airline depend on the contribution and co-operation of all the parts of its organisation. Part of achieving full co-operation is for each part of the organisation to understand the contribution the other parts make to the airline's efficient operation.

Essentially, an airline's finance and accounting functions support the airline with two major resources; firstly, the provision of financial information, by reporting to an airline's owners and management the financial results and the progress towards achieving the airline's long-term financial goals. The reporting is to each part of the airline and to third parties. Secondly, the finance function is responsible for managing one of the airline's vital resources, its finances. This description is simplified, as the real world is more complicated than that. It is helpful for all airline staff to spend some time finding out how accounting and finance help the airline to succeed and how they help all of the airline's staff to do their jobs.

This book will help an airline's operating staff understand what the finance and accounting function is responsible for, how it operates, how it is organised and what its relationships are. The objective is not to turn the reader into an accounting or financial expert, but to describe the essentials of their functions and give everyone a broad understanding of how finance works.

Appreciating the essentials of reporting, financial controls and financing will help operating staff make decisions which are consistent with the airline's financial and operating objectives and, by doing so, help the airline be successful. Improved understanding will help co-operation within the company. If a team wants to be a success, each member needs to understand what the other players do, as well as how and why they do it.

An airline has specialist departments, each of which has its own technical words, phrases and jargon, which are not always easy to understand. The accounting and finance function is not an exception to the rule; it has its own 'language'. The most common financial jargon used in an airline is explained in the book.

The explanations of the 'why' and 'how' of finance, together with the explanations of its jargon, will remove any confusion about the contribution the finance function makes to an airline's success.

1 The Finance Department

What it does and how it fits into an airline's organisation

Before explaining the details of the department, it is important to make clear that there is no one ideal way of organising an airline's accounting and finance functions, as each airline will have its own preferences for the way its management is organised and responsibilities are allocated. Many airlines combine the accounting and finance operations into one department under the leadership of one person, typically titled Finance Director or Chief Financial Officer (CFO), who is usually, but not always, a member of the Board of Directors. A combined department, i.e., accounting and finance, is a popular option and, in this book, references to 'Finance Department' assume a combined department.

The role of the Finance Department

The Finance Department provides services to the whole airline. The basic services are recording and reporting financial information and managing the airline's financial position. The financial information held by the Finance Department and the reports issued, both routine and ad hoc, help managers record progress towards meeting the airline's objectives and to make decisions which are consistent with those objectives. Information is a resource. Imagine that the data sent to the Finance Department is its raw material and the reports they issue are its finished product. The real trick is for the reports to provide the right numbers accurately for managers, and this requires co-operation and communication between the Finance Department and all other departments. Within the airline there should be regular reviews of all routine reports, including those issued by the Finance Department. The review should cover whether the reports are still needed, and their content and layout, the frequency of issues and how quickly after a period end they are issued. Circumstances change, and it is important that all reports, not only financial reports, are relevant and help managers do their jobs. There isn't a part of an airline that is not touched by finance. The financial information held in the Finance Department's records is a resource for the whole airline.

Organisation and responsibilities of a Finance Department

A Finance Department can be organised in different ways and the structure depends on the airline's management approach (see Figure 1.1). A traditional organisation is to have a separate section for each of the major operations of the department; in this book, the most common titles of each section have been used, but they may differ by airline.

With this approach the responsibilities of the typical sections are:

- **Statutory Accounting:** production of the financial reports required by law; these are discussed in Chapter 8. Included under this heading is the keeping of a record of all the airline's tangible and intangible assets, the original cost, the cost of any additions or modifications, an estimate of the assets' useful life in the airline and its overall useful life, together with a periodic assessment of the current value and some means of identification. The estimates and assessments should be reviewed regularly.
- **Management Accounting:** producing reports for the airline's management, budgets, forecasts and actual results; this is the area where most operating staff have contact with the department and are most likely to use the information provided.
- **Revenue Accounting:** recording the airline's actual revenue; this work can be quite simple or very complex depending on the airline's business model. The simplest records are for an airline that only operates within one country:

  - with all its tickets sold on its own website
  - passengers paying in advance
  - does not make refunds if tickets are not used
  - its tickets are only to be used on its own flights.

  In this case, the records only need to identify the passenger, the flight and the amount paid. The amount received in advance will be recorded in

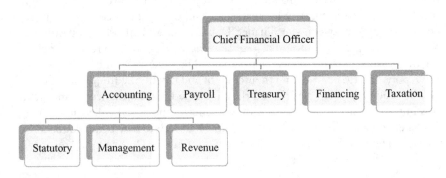

Figure 1.1 Finance Department's functions

an account named something like 'Unearned Transportation Revenue'. When the passenger travels, the amount received from the passenger will be transferred from the 'Unearned Transportation Revenue' account to the airline's 'Profit and Loss' account as revenue.

Generally, the more options and services an airline offers, the more complicated the revenue records become; if, for example, in addition to the services offered in the example above, the airline:

- sells its tickets through its own offices, travel agencies in different countries, as well as through its own website
- issues tickets which may include flights on other airlines
- accepts another airline's tickets on its flights
- permits refunds of unused tickets
- offers packages which include a flight and hotel stay, with or without tours.

The records become very complicated, more records are needed and there are calculations for such things as the proportion of a partly used ticket which should be refunded. The accounting effort required by the more complicated business model is significantly larger than for the 'simple' one.

The accounting for revenue can be further complicated if the airline has a Loyalty programme, particularly if the airline 'sells' miles to other service suppliers (e.g., car rental companies). The amounts involved in Loyalty programmes can be very substantial and many programmes have been transferred into separate operating companies.

Revenue records are usually kept on the airline's computer records.

- **Payroll:** calculating and arranging payment of staff salaries; frequently this section is located in a separate area due to privacy concerns.
- **Treasury:** managing and forecasting the 'cash-flow', which is the money received and paid out. This is an important area because the company is required to be 'solvent'; that is, able to pay its debts when they become due. Managing an airline's cash-flow is an important topic involving many operations, including such matters as investing surplus funds overnight and, in the longer term, also handling the airline's foreign currency position. This is discussed in Chapter 6.
- **Financing:** managing the airline's financial position; that is, its borrowings and financial risks. These are discussed in detail in Chapter 4.
- **Taxation**, preparing all 'tax returns', i.e., the calculation of any taxes due to be paid or refunded; an airline's tax position can be complicated if taxes are payable in many countries. The airline's foreign tax position in a foreign country may be simplified if a Treaty for the Avoidance of Double Taxation, often referred to as 'Double Tax

Treaty/Agreement' or 'DTA' exists between the two countries, but all foreign taxes need to be monitored.

These are the basics, and it may be that the Finance Department will be asked to take on other responsibilities, such as negotiating and dealing with insurance matters or issuing invoices for non-travel income, e.g., aircraft handling.

On the surface it may seem that some of these operations, like paying suppliers, would be better done within the various operating departments in the airline. In general, the reason for splitting the responsibility for some processes (e.g., making payments) between different departments is part of the company's system of 'checks and balances' (i.e., splitting a responsibility reduces the chances of fraud). The operating department is responsible for reviewing, checking and approving each supplier's invoice and the Finance Department will check that the invoice has been correctly approved before arranging payment.

Location of the finance staff

A Finance Department provides services to all parts of an organisation. The department may be 'centralised', i.e., all sections are located together, or 'decentralised', or a combination of both. A popular option is locating part of the finance team with the major operating departments. This has the advantages of improving communication and understanding between departments and the finance staff. A possible disadvantage is that the co-operation can reduce the effectiveness of the airline's checks and balances. Also, the finance staff may feel isolated from the main Finance Department. The latter problem can be overcome by good communication between the departmental finance team and the corporate finance team. It is not unusual in the real world for the Finance Department's staff to be partly decentralised to one or two large departments (e.g., Engineering) and partly centralised, depending on the preference of the operating department's manager. Even if the department is decentralised completely or just to some extent, the staff are still part of the Finance Department, which remains responsible for the quality of work, discipline and remuneration.

Operating structure

The Finance Department has dealings with many parts of an airline's operation. Figure 1.2 presents a diagrammatic overview of these internal connections.

The Finance Director or CFO usually reports to the Chief Executive Officer, or whatever the senior operating position in the company is titled, but in some cases they may report to an Executive Chairman. In addition, there are two sub-committees of the Board of Directors which influence the work of the Finance Department. These are the Audit Committee and Finance Committee.

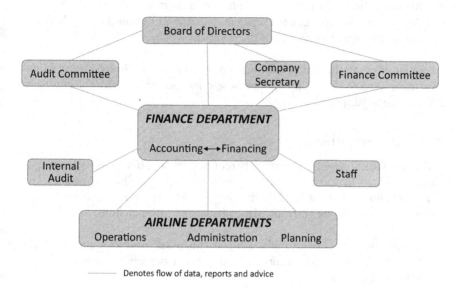

Figure 1.2 Connections inside the airline

Audit Committee

Although having an Audit Committee is not a legal requirement in every country, it is recommended as part of good 'Corporate Governance', which concerns the overall management of a company. If it is a legal requirement, the law will define the committee's duties. If the airline's shares are quoted on a stock exchange (i.e., it is a 'Public Company'), it is almost certain that a stock exchange will require the company to have an Audit Committee. The Board of Directors has the power to allocate additional duties to this sub-committee as it sees fit. The usual duties of the Audit Committee are to:

- Establish that the company's accounting policies are appropriate; these will be discussed in Chapter 8.
- Ensure the company's financial reporting is complete and complies with current 'Reporting Standards'; these are national (or international) accounting and reporting standards applicable to all companies.
- Monitor the airline's 'internal controls', which are the systems designed to prevent fraud.
- Ensure compliance with the airline's financial policies (e.g., regarding the treatment of Debtors and Creditors).
- Identify and monitor the ways the airline's risks are managed.
- Choose and monitor the performance of the 'External Auditors', who are independent accountants employed to review and confirm the accuracy of the airline's 'Statutory Reports' (i.e., the reports required by law; these are further discussed in Chapter 8).

- Monitor the performance, work plan and reports of the airline's 'Internal Audit Department'; this is a separate department responsible for checking the accuracy and efficiency of the company's systems.

The members of the Audit Committee should all be 'Independent Non-Executive Directors' (i.e., directors whose only connection with the company is their directorship).

Finance Committee

A Finance Committee is not currently required by law, but most large companies and airlines have one to help manage the company's financial position. Its duties are allocated and agreed by the Board of Directors. Hence there can be a variation company by company. In general, the duties include:

- Monitoring the company's general financial position and 'Capital Structure'; this is the company's equity and borrowing position and is discussed in Chapter 5.
- The timing and sources of fundraising.
- Reviewing 'financial exposures'; these are discussed in Chapter 4.
- Reviewing the reporting of the airline's financial position.
- Monitoring financial relationships; these are entities to which the company owes money, or which owe the company money, together with the amounts owed.

The committee usually consists of some Executive Directors and Independent Non-Executive Directors, together with senior staff from the Treasury and Financing functions. Like all committees, the operation of this committee should be reviewed by the Board of Directors regularly, (say) every three years to ensure its duties and membership are appropriate.

Because both the Audit Committee and Finance Committee are sub-committees of the Board of Directors, the minutes of their meetings are circulated to every member of the Board. The Board of Directors remains ultimately responsible for the decisions made by each sub-committee.

The Finance Department has significant internal relationships with the Internal Audit Department and the Company Secretary. Both of these functions are independent of the Finance Department, but the work of each of the three departments is closely related. Frequently these two departments are seen as backroom administration, but they can have a significant influence in their own way on the operation of the whole airline.

Internal Audit

The Internal Audit Department frequently reports to the Chairman or Chief Executive Officer in order to ensure its independence. The department does not make decisions, but is an internal advisor with a main responsibility to review and assess the company's reporting, control, risk management and governance practices and procedures to ensure they are effective, and where they are not, to recommend improvements. The department's focus is on how to operate more effectively. There is not a legal requirement for a company to have an Internal Audit Department, so the department's responsibilities are usually defined by the Board of Directors based on its experience and a recommendation of the Audit Committee.

The department's 12-month work programme is usually approved by the Audit Committee and the department's reports and recommendations are issued to the Audit Committee with copies to the operating departments mentioned in the report. A regular report is submitted to the Board of Directors summarising the Internal Audit Department's findings, the recommendations made and progress implementing them. There are occasions when the department's recommendations are not accepted by the operating departments affected, and when this happens the Board of Directors must resolve the matter. The company's external auditors usually have access to the Internal Audit Department's reports. The external auditors may also recommend or suggest to the Audit Committee areas of the airline's operations that may warrant a review. It is important that the internal and external auditors do not duplicate their work. The responsibilities of the internal and external auditors differ but they are complementary.

Company Secretary

It is difficult to produce a general list of responsibilities for a Company Secretary. There is a great deal of country-by-country variation in such a list. Some countries do not require small companies to have a Company Secretary. At the most basic level, the duties of a Company Secretary are to ensure the company complies with the local Companies Act by submitting required reports to government departments. These are such reports as a list of the company's directors, and keeping required records of matters like shareholders' details and legal charges on the company's assets. All of the 'required' items will be specified in the Companies Act. Where there is no requirement for a company to have a Company Secretary, the Board of Directors will be responsible for submitting the required reports and ensuring the required records are kept.

Depending on the country, the additional duties which may fall to the Company Secretary are:

- Advising directors of their legal and corporate responsibilities and the penalties for not doing them.
- Advising the board of directors on corporate governance matters (i.e., how the company is organised and managed).
- Organising shareholders' general meetings, producing the agenda and minutes.
- Assisting the company chairman to prepare the agenda for meetings of the Board of Directors, as well as producing the minutes of meetings.

- Providing general legal advice to the Board of Directors.
- Advise on and organise corporate governance training for directors.
- Obtain any additional information when requested by directors.
- Be responsible for the storage and security of the company's essential documents in addition to the statutory requirements (e.g., property title deeds, insurance policies).
- Arrange dividend payments to shareholders.

The CFO and Company Secretary work closely together in producing the company's reports to shareholders.

Although the Company Secretary is an independent function, in some smaller companies the CFO may also act as the Company Secretary. Where the responsibilities allocated to the Company Secretary are purely administrative and routine, they may be sub-contracted to a third-party company which specialises in this type of work.

Information and provision of services

In order to do its job of producing statutory reports and providing information and services to the airline's departments, the Finance Department needs a flow of information from the rest of the organisation to add to the information it already has. Below are a couple of examples which illustrate this interdependence, as well as showing how information is brought together to be used by managers.

Producing a monthly Profit and Loss Report

This report explains what has happened and will show the total figures for the whole company supported by the results for each business activity, as well as the income and costs of each department. This report is one of the reports the Board of Directors uses to help them assess whether the airline is on track to meet its short-term financial goals. The supporting information on business activities and departmental income and costs helps managers evaluate how each business activity is performing and whether the departments are likely to meet their objectives.

The basic information needed for this report is derived from,

- Finance Department
  - Revenue earned from the carriage of passengers and cargo
  - Interest earned
  - Payroll costs
  - Interest payable
  - Depreciation
- Each operating department
  - Ancillary income earned (e.g., check-in services for other airlines)
  - Aircraft operating costs incurred

- Administration costs incurred
- Stocks consumed

Here 'earned' means the amount due from a third party for a completed transaction regardless of whether the money has been received, and 'incurred' means costs which the airline has committed to even if they have not been invoiced to the airline and paid.

A great deal of information is used to produce these reports. The airline's Finance system needs to collect and record it efficiently. Designing the main report and supporting reports so they can be quickly and easily understood is important or directors and managers will be overwhelmed by data.

Producing a plan for the airline's cash position

This report will generally be prepared by the Finance Department and deals with what the airline's funding position is likely to be in the future. It is usually used by the board of directors partly as information on the solvency of the company and partly for planning major changes in the airline's finances, for example, whether to list the airline's shares on a stock exchange. The Finance Department also uses the information to help it plan the financing of individual assets, the investment of short-term and long-term surplus funds and to ensure the airline is able to pay its debts.

To produce the report the Finance Department needs to know what the airline's expected revenue will be, what orders are being placed or are expected to be placed for goods, services and assets, and when the costs are likely to be paid.

The report is a forecast that, almost by definition, will not be completely accurate because it is based on a combination of current information, assumptions and historic trends. The amount of detail needed to prepare the report and to be included in the report needs to be decided and a balance struck between the effort required to generate the basic information and the accuracy of the final result. This is also discussed in Chapter 6.

The basic information will come from each operating department and is then fed into the department's system to produce a forecast. Depending on the airline's systems, much of the information will be handled by computerised systems.

Producing accurate and useful reports requires a good flow of data into the Finance Department. In addition to the routine information on amounts to be paid to, and amounts to be collected from third parties, there is other information which is essential for the work of the Finance Department. Examples of key information are:

- Forecasts of revenue; these are usually provided by Revenue Management or the Planning Department.
- Plans for aircraft purchases and disposals; these will come from the Planning Department with significant input from, for example, the Engineering and Flight Operations Departments.

- Plans for major advertising campaigns which will come from the Marketing Department.
- Changes in the amount held in 'Stocks' (i.e., supplies held); this can either be an increase or decrease and can be from almost any department, but the majority of high-value stocks are controlled by Engineering (e.g., engines, rotable parts, spares). Inflight Services (e.g., crockery, cutlery, napery), and Marketing (e.g., promotional and advertising material like pens and T-shirts) tend to be low-value items but high in volume.

Nevertheless, each operating department is responsible for producing comprehensive input, just as the Finance Department is responsible for producing a realistic result. The airline is one complete unit even though it is divided into departments. Co-operation, co-ordination and assistance between departments are vital. The better the flow of information is, the better the reports and forecasts produced by the Finance Department will be and, accordingly, the control of the airline's money will be improved.

Workflow

The workflow in the Finance Department is basically according to a monthly cycle. For example, creditors' invoices are usually paid once a month, staff may be paid monthly or more frequently, and reports of revenue and expenditure are usually issued to operating departments monthly, and ad hoc reports will always be required.

There is, however, a longer work cycle for the preparation of statutory reports. The format of statutory reports is usually governed by reporting standards, and although the information in the airline's management information system is the same as in the statutory reports, the layout of the reports can be very different. The preparation of these reports produces an increase in work, partly because of the format of the information required to be reported and partly because the company's external auditors are required to make an 'audit'; that is, a thorough inspection of the reports because they are required to express an opinion as to whether the reports show 'a true and fair view' of the company's financial result and financial position.

External influences on reports

There are changes in progress to a company's statutory reporting which will require the inclusion of more information, some of which is not financial, and this requires the assistance of other departments; this matter is discussed in Chapter 9. The requirement to issue statutory reports varies between countries. The minimum requirement is to produce a full set of reports once a year for the year. If the airline is a public company, the country's 'stock exchange' will probably require additional reports. For example, a simpler 'Interim Results' set of reports may be required halfway through the year. Some companies are required to issue a report of their results after the first and third quarters as well. Hence, in this case, the Finance Department may be required to produce some form of statutory report each quarter. In addition, it will probably be necessary to issue a special report when there are exceptional circumstances, such as, for example, an unexpected rise or fall in profits.

When producing the statutory reports, the Finance Department liaises closely with the airline's Company Secretary who is responsible for ensuring that the company complies with all legal requirements, including the regulations of any supervisory authority, e.g., a stock exchange, and the company's own regulations. The Company Secretary also advises the directors to ensure they understand their legal responsibilities relating to the statutory reports.

Staffing

There is usually a mixture of staff and skills in a Finance Department. The team will probably consist of professionally qualified accountants, qualified treasury professionals, staff studying for an accountancy or treasury qualification (sometimes both) and staff with experience in the department and in other areas of the airline. It is not unusual for Finance Department staff to transfer to other departments after a few years' experience because their knowledge of accounting and/or finance is seen as a valuable asset by other departments. Similarly, staff transferred to the Finance Department from operating departments bring with them useful experience of how the airline actually operates.

The Finance Department provides financial information to users across the whole airline, using its own information and incorporating information from all other departments. Hence all managers in the airline need a comprehensive understanding of the needs and responsibilities of the Finance Department, just as the Finance Department needs to understand the airline's operation.

2 Financial strategy

Why a financial strategy is needed, what it looks like and how it is decided

Every airline has a view of its future. Companies can use different words or terms to describe this view; it may be called a 'Strategy' or 'Vision' or 'Target'. This long-term view must cover all parts of the business and finance is one of these parts. This chapter deals with the financial part of an airline's strategy. The overall strategy defines what the airline is seeking to achieve and is an important working document. The financial strategy defines what the financial part of the business should look like in the future.

The 'financial strategy' is an agreed opinion as to how the airline's long-term plan will be financed and defines what an airline's financial position will look like if the strategy is successful. The financial strategy is part of the airline's overall long-term plan and sits beside such items as Customer Satisfaction, Safety, Environment, etc. It is likely that the Finance Department will be involved in the discussions of all the headings in the general strategy.

The general sentiment of a financial strategy is often something like 'The airline will have a strong financial position; it will be able to reward shareholders with dividends and staff with good salaries; it will be able to raise funds at a reasonable cost; and it will successfully manage its financial risks'. The intention is clear, but for the strategy to be put into practice and become a guide for decisions, words like 'strong' and 'reasonable cost' have to be defined and, in some cases, values given to them. Once there are definitions, the strategy will be clearer, and can be more easily understood and provide a guide for decisions. The strategy and definitions should be worded so that in normal circumstances it does not need to be changed. Yet it must be possible to review and, if necessary, amend it if circumstances change significantly.

Finance framework

Within the airline's overall strategy, the finance section of the strategy defines what the airline's financial future will look like. Meeting the requirements of the financial strategy requires the co-operation of the whole airline and influences the decisions made in operating departments. In particular, the financial strategy provides the Finance Department with a framework for its operations and guidelines for the transactions of the Treasury and Financing functions. The practical use of a financial strategy is to:

- give a framework for management's decisions
- highlight changes that need to be made
- help set priorities
- identify the resources needed
- provide a basis for measuring the financial performance of the airline
- help the airline adapt to changing circumstances.

Without a strategy for finance, the airline will lack direction and will not have a reference point to use when seeking to resolve problems or when deciding which financing option to take. A ship without a rudder will simply follow the wind and the waves. For example, if the best financing option being offered to the airline will result in (say) more than 60% of the airline's assets being used as security, thereby exceeding the level permitted in the agreed financial strategy, the offer will either have to be rejected or re-negotiated, or the airline will have to raise unsecured debt, or issue more shares, or some combination of these options. A clearly defined financial strategy will help the staff in all departments understand clearly what management is trying to achieve.

It is probably easier to understand a financial strategy by looking at one for a hypothetical airline. The parts in the financial strategy will depend partly on the airline's current financial position, but mostly on its desired future position. There are likely to be circumstances where the current financial position is not the desired position, in which case the strategy should be supported by a plan as to how to improve from the current position. This is the financial strategy of a fictitious airline as shown in Table 2.1.

Table 2.1 Financial strategy of a fictitious airline

Element	Measure		
Not more than:			
Gearing	Debt	60%	not to exceed 70%
	Equity	40%	
Security	Maximum	60%	of asset value
All-in cost of debt		1%	over U.S. Treasury Bonds
Debt servicing		70%	of cash-flow
Not less than:			
Reserve funds	Minimum 3 months		of operating costs
Profitability	Minimum	2.5%	of revenue
Return on Capital Employed	Minimum	4.5%	of the equity ad loan capital

Understanding terms

The expression of the various guidelines included in the financial strategy should be understandable by everyone and explained to all the airline's staff, because the implementation of the financial strategy will affect all of the airline's operations, and influences the money available to operating departments and for investment in assets like aircraft and stocks.

The meanings of the various parts of the strategy are as follows:

Gearing

This is the ratio between the money borrowed from third parties such as banks (this is the Debt) and the money the shareholders have invested in the airline plus the profits which have not been paid out as dividends to shareholders (this total is the 'equity', sometimes referred to as 'Shareholders' Funds'). The ratio may also be called the 'debt/equity ratio' or the 'gearing ratio'. If the airline is a public company, the equity is the amount of money actually paid to the airline by the shareholders, not the current market value of the airline's shares.

Lenders and investors use the debt/equity ratio to judge whether a company has too much debt compared to its equity. Excessive debt is assumed to increase the risk of the company failing to pay the interest due and/or repayments due on its debts and, in the most serious cases, failing as an operation and ceasing to exist. Whether the total amount of debt is excessive is a matter of opinion and can change when circumstances change. This makes it essential for the airline to monitor its debt and equity position frequently, and to be aware of the current market view on debt and financial risk. Frequently a useful way to get advice on a good capital structure is to look at two 'markets' for guidance;

- The range of debt/equity ratios, and other ratios which the airline's bankers and industry bankers and analysts consider to be acceptable, often termed as 'sound'.
- Other companies in the airline's home country that are thought of as being financially sound to examine their debt/equity and other ratios.

There are various ways of expressing the debt/equity ratio; usually it is 'times', which is debt divided by equity. In Table 2.1, the figures are debt of 60 divided by equity of 40, equalling 1.5 times. Although many investors do not look beyond the debt/equity ratio, it is advisable to apply other tests to see whether a company has borrowed too much. An additional test is to calculate the company's 'interest cover' which compares the airline's 'operating profit' (profit before charging interest and tax) also referred to as EBIT, with the interest paid on loans. The cover is usually expressed as 'times covered' and is a straightforward calculation. For example, if the operating profit (EBIT) is USD 20,000,000 and interest is USD 4,000,000, the interest is five times covered. What is a good figure for interest cover depends on the industry the company operates in and how the company compares with its peers. Clearly, the lower the figure for interest cover the more concern lenders will have, and there is a general view that 2.5 times or lower is a cause for concern. Although there are other measures to assess the level of debt, these are the most common.

In Table 2.1, the note 'Not to exceed 70%' adds some flexibility to the target, but also establishes an absolute maximum for the amount that can be borrowed. The 70% limit acknowledges that borrowings are likely to be made irregularly. For example, if there is a substantial increase in the aircraft fleet in a short period, total debt may temporarily exceed the 60% desired level for a time. However, if the increase in the fleet will push total borrowings beyond 70%, then to meet the airline's objectives and stay within the definition in the strategy, other methods of financing the purchases will need to be found, such as issuing more shares.

Deciding on the correct or acceptable or target debt/equity ratio is not easy because a major influence is the attitude of investors towards investing in airlines and whether lenders want to lend to airlines. These views are not fixed and are influenced, inter alia, by the expected profitability for aviation generally, as well as the current view of the financial prospects for an individual airline. The airline's discussion will take into account:

- the debt/equity ratios of comparable airlines and probably other ratios like interest cover
- what lenders think is an acceptable ratio
- the likelihood of the need to borrow substantially in the near future
- the possibility of raising equity by issuing shares.

The decision involves weighing opinions and intangible factors.

The definition of what is included as 'debt' in the debt/equity ratio is likely to change with the introduction of the requirement in International Financial Reporting Standard 16, which expresses that operating leases be recorded on a company's balance sheet.

Security

This is the proportion of the airline's assets which is permitted to be used as security for amounts borrowed. Setting the amount requires serious thought, and it is often difficult to reach a decision because opinions vary. Traditionally, banks and other lenders have been interested in lending to airlines against the security of aircraft, because it is believed that aircraft tend to hold their value even during difficult trading periods in aviation, hence they are considered to be high-quality security. If an airline has all or almost all its assets pledged to lenders, it has a risk of getting into financial difficulties during times when it is difficult to make profits, hence generate cash. Having a limit on the value of assets that can be used as security for borrowings can help to reduce this risk. If there are 'free' assets which are not being used as security for loans, these assets can be used in an emergency as security for an advance of funds, either for the long-term or the short-term.

All-in cost of debt

This refers to the annual total cost of all borrowings. It will consist mainly of interest but should include any fees and maintenance costs solely related to borrowing as well. An example of a maintenance cost related to borrowing would be if the lender requires that the security for a loan should be held in a separate trust. The cost of running the trust is considered to be part of the cost of borrowing.

Frequently, the rate of interest payable on debt is expressed as a 'margin', which is a percentage above a 'benchmark rate', which is a published rate of interest. Examples are '10-year U.S. Treasury Bonds' and 'LIBOR' meaning 'London Interbank Offered Rate', but there are others. Even if the rate of interest for a borrowing is fixed for the whole period of a loan, the interest to be charged is likely to be expressed as a margin over a benchmark rate and will be applied using the benchmark rate on the day the loan is made to the airline. The discussion of the maximum an airline is prepared to pay in interest and fees will take into account a lot of factors, including:

- What other comparable airlines pay; it is very difficult to establish how much other companies pay for their debt as the information is usually considered to be confidential. In practice, an estimate can be made by using a careful examination of the figures in the airline's statutory reports.
- What other companies pay. Similarly, this information is not readily available and will probably need to be estimated.
- The current view of lenders as to whether they wish to lend to airlines. This guidance can be established by speaking to current and potential lenders. Lenders, as part of their risk management, establish limits on how much they will advance to each industry and each company within an industry.

Debt servicing

This is the amount of money required to pay the interest due and any loan repayments due each year.

Establishing a limit for this to a certain percentage of the airline's cash-flow is part of ensuring that it can meet its obligations as they fall due, even during difficult trading times. It is part of developing an approach which will produce a strong financial position and a good reputation.

Reserve funds

This is the amount held to meet cash shortages in difficult times or in emergencies. There is not a generally accepted formula for calculating how much an airline should hold to deal with emergencies or short-term trading difficulties. Two common approaches are to hold a number of months, often three (for no particular reason), of operating costs or a number of months of debt servicing costs, including lease payments.

Solvency

These last two decisions are involved in the important topic of 'solvency', which is the requirement for a company to be able to pay its debts to suppliers, including staff and lenders when they are due to be paid, and is a vital consideration for any company. Sometimes solvency is assumed to be the ability to repay all borrowings at any time, whether they are or due to be repaid or not, but this is not the case. An airline is solvent if it can pay all the amounts that are currently due to be paid and is forecast to meet all its obligations in the future.

In addition to the practical problems of not being solvent (i.e., the problem of dealing with all creditors), there are usually legal consequences (for example, directors may be held legally responsible) if a company continues to operate whilst insolvent. The discussion on debt servicing and reserve funds generally focuses on how much money the airline should have available to cover short-term cash shortages. An airline's major cash needs tend to arise irregularly, and the concept of prudence requires that some cash be held in reserve for 'rainy days' to ensure all obligations can be met on time. If there is a short-term cash problem, the value of the airline's assets does not help unless the assets can be sold quickly; rather, what is important is how much cash the airline has available. How the reserve funds can be invested is discussed in Chapter 6, but generally they will be invested in low-risk, short-term deposits or bonds. The funds must be available at short notice. Restricting the value of assets that can be used to secure borrowing also helps to support an airline's solvency by having assets available to support borrowings if funds are needed quickly, provided the borrowing does not breach the required debt/equity ratio.

Sources of funds

There are only three main sources of funding for a company;

- equity, which is selling part of the ownership of the company for money (e.g., shares)
- borrowing money from a third party
- generating funds internally by making profits from business operations, selling assets and minimising the funds used in the business

The third is perhaps the most important.

Profitability

The reason for including profitability in the financial strategy is because it is, or should be, the regular source of funds to operate and expand the business. An airline needs to consistently make profits to remain solvent and develop. Profits and cash are not the same thing, and this is discussed in Chapter 5. Without profits the airline cannot be financially sound or strong. The problem when deciding on the financial strategy is what minimum level of profits should be set and how should it be expressed. Aviation is highly competitive, and profits are not guaranteed. Establishing a minimum level of profitability, such as the case of 2.5% in Table 2.1 above, helps to guide management when actual and forecast profits are discussed. If profits are forecast to drop below the required minimum then some sort of remedial action will definitely need to be taken. Setting a minimum level of profitability does not mean that the management should not seek to achieve a higher level, but sets a warning level that action is needed.

The example in Table 2.1 uses the airline's profit after tax as a percentage of the airline's revenue as its indicator of profitability. In simple terms, to meet this requirement means that the airline must increase its profits as its revenue grows. This is a common way to express the profitability of any company, but it is not the only way. It could be that the airline's management believes that a better measure of profitability is the 'return on equity', which is the profit expressed as a percentage of the amount shareholders have paid into the airline plus the profits which have not been paid to shareholders as dividends, for determining the minimum level of profit, the argument here being that a target return based on the amount shareholders have invested better measures how efficient the company is being run, and permits comparison with other investments. There is not a standard way to calculate the figures used in the return on capital, but often the way in which the company calculates the figures is shown in their statutory report.

Management may believe that the figure for the 'return on capital employed', which includes both equity capital and borrowings, is a better measure of the way management are using the funds at their disposal.

Not every airline's management believe that the key measure of profitability should be based purely on financial figures. The management of a low-cost airline, for example, may think it more useful to relate profits to the number of passengers carried. Other airlines will prefer to use more than one ratio to cover profitability.

Measuring profitability is a very difficult subject as each ratio has its advantages and disadvantages.

The key issue here is that a measure for a minimum profit level should be set and used consistently. Measuring profits will be discussed in Chapter 7.

Other areas

There is not a limit to the items which can be included in the airline's financial strategy, but in general it often best to try to restrict the items mentioned in the strategy to those which are considered to be essential. Examples of additional items that may be included are the level of dividends to be paid to shareholders, the maximum levels of stocks to be held and the criteria for investing in other companies.

Interpreting and developing the strategy

The measures in the financial strategy tend to state maximum limits which should not be exceeded (e.g., the debt/equity ratio), but others are minimum levels which must be achieved or improved upon (e.g., profitability). This is why the strategy is split into two parts: 'not more than' and 'not less than'. Once the details of the airline's financial strategy have been agreed, it is not to be 'filed and forgotten'. Whenever the airline's actual financial results or financial forecasts are produced the figures will be compared to the strategy. When major decisions are considered, part of the financial considerations will be the probable effects on the airline's financial position and compared with the desired position as expressed in the financial strategy; the strategy is a definition of what the airline seeks to achieve.

Most often a strategy is developed once an airline has been operating for some time. Despite the importance and usefulness of a financial strategy, it is unusual for a newly formed airline to have a strategy from the first day, often because the airline starts on a small scale and the pressure is to start operating. This means that a strategy will often be developed by an operating airline, and achieving the measures in the financial and other strategies will almost inevitably require some parts of the airline's organisation to change the way they work. Changing the way things are done is not always easy and time is needed.

The first time the financial strategy is developed it is possible that some parts of the airline's current financial position will be outside the limits set. In this case, a plan needs to be developed and agreed upon for steps to improve the position. The plan will usually be recommended by the Finance Committee based on a recommendation produced by the Finance Department. An example of the sort of change needed is: if the airline's reserve funds are less than the desired level, it may be decided to reduce the amount in stocks held or to reduce the airline's debtors to generate funds, which can be used to boost the reserve fund to the desired level. It is likely that bringing the airline's current financial position into line with the strategy will require the co-operation of all the airline's departments. Similarly, assisting other departments to meet their strategic goals may require the Finance Department to change the way it works or to produce a different form of report. Implementing the plan could take a year or more because the financial strategy will have an element of targeting in it.

The financial strategy is approved by the airline's Board of Directors. In practice, the research needed to reach a decision on the detail of the strategy is developed by the Finance Department, probably in consultation with the airline's leading lenders and with references to other airlines' financial positions. It will also be important to consult other operating departments. A recommendation or, even better, a recommendation with some options, will be prepared by the Finance Committee for the Board of Directors. The financial strategy should not be developed without taking into account the other areas of the airline's strategy because all the other areas involve finance to some extent.

Reviewing the detail of the strategy

Just as the financial strategy is an active document, the details of the strategy must stay under active review. Once it has been agreed, a financial strategy is not set in concrete; it is more like it is set in 'wet' concrete – it can be changed if there are major changes in circumstances, not just in the airline but also in such areas as financial markets and the world economy. Examples of major events which would prompt a review of the financial strategy are: a strong desire to accelerate the delivery of new aircraft to meet expected market demand, or a substantial change in benchmark interest rates or the availability of funds to borrow. Amending the detail of the financial strategy should be done cautiously. The airline's management may take the approach that the changes in circumstances are temporary and should be treated as short-term deviations from the basic financial strategy. If this is the case, the management will recommend revised temporary targets for the expected duration of the deviation to the Finance Committee for review and to the Board of Directors for approval. The important point is that any temporary deviation from the strategy must be planned and controlled. It might be tempting to make the strategic targets that are easy to achieve, or to change the detail to ensure a particular decision can be made, but any pressure to do this should be resisted because it may be detrimental to the airline's long-term financial health and, in extremis, its continued existence.

The process of reviewing the detail of the financial strategy to see whether any element should be changed is similar to developing the original strategy and is usually initiated by the Finance Department. This does not mean that only the Finance Department can request a review. The strategy is a company document and decision; it is open to any director to suggest that a review and possible change is appropriate. A recommendation to amend the items in the strategy or add new items should be prepared within the Finance Department and will include an assessment of the effect of changing or not changing the item. It is almost certain that one of the assessments will be the effect on the airline's reputation, in particular amongst the financial community, but also by the general public. The recommendation should be examined by the Finance Committee and passed to the Board of Directors for discussion and possible approval. The following two examples illustrate the circumstances which may generate changes in the detail of the financial strategy. The first is a change generated by a technical accounting issue and the second is following a change in the way the business operates.

Example 1. International Financial Reporting Standard 16 requires that operating leases be recorded in the company's accounts as assets and obligations.

The current practice is generally to detail a company's operating leases in a note to its statutory accounts. The likely effect of this reporting change is that a company's liabilities will increase. Depending on the value of operating leases a company has and the presentation of the new information in the statutory accounts, the company's debt/equity ratio may change. If there is a significant change the company's management will need to consider how to deal with the change. Clearly, the company's actual circumstances and liabilities have not changed, but the presentation of information has. The most likely solution is for the obligation for operating leases to be shown separately in the financial strategy under 'Not more than' with its own limit, and the limit will be approved by the directors. In the statutory accounts there will be a detailed note explaining the reason and effect of the change if the figure is significant.

Example 2. An airline may decide that its level of customer service would improve and it would generate additional profits if its check-in and despatch services at its home airport were performed by a separate company of which the airline owns 75%. In addition, the new company would offer its services to other airlines. This is a significant change. Each one of the definitions in the financial and customer service strategies will need to be reviewed to establish to what extent the figures need to be amended.

The new company will probably need additional assets and these may have to be financed, which in turn might affect the airline's gearing. A new minimum level of profitability for the remaining part of the airline may be needed. In addition to these possible changes to the airline's financial and other strategies, an entirely separate set of strategies will be needed for the new company.

The strategies for each part of an airline's operations must be understood by the whole organisation because they affect the whole airline. Just as the Finance Department is involved in other aspects of strategy like customer service, other departments are involved in the financial strategy. If there is not unity of purpose not only will the organisation not work as a team, but worse, some staff may make decisions that actually make it more difficult to achieve the financial and other strategies.

It may be that the initial thought is that the financial and other strategies are commercially sensitive and hence confidential documents that must be kept secret; after all, they define what the airline is trying to achieve. Further thought shows this cannot be the case; the strategies must be understood by all the airline's staff because they have to work to achieve them. It is not easy for an airline's management to bridge the gap between the commercial confidentiality of its strategy and the need for all the airline's staff to understand its goals. The problem becomes more difficult for an airline with shares quoted on a stock exchange depending on the exchange's regulations. Companies generally are not required to reveal information which is commercially sensitive, but if a public company tells all its staff, there is the question of whether it should also tell current and potential investors. Because communication with staff is so important, the bridge is frequently producing a statement of strategies which clearly gives the spirit of each part of the strategy, but only includes detail which is not commercially sensitive. Some airlines publish all or some part of their strategies in their annual reports.

When reading a financial strategy, it may appear to be a scientifically structured statement, but the discussion in this chapter has shown that, although the strategy is generally expressed in numbers, actually supporting the numbers are a whole range of opinions and assumptions. This part of the airline's strategy is an expression of the Board of Director's view of what the airline's financial position should be so that it is financially secure and is seen as being well managed financially.

3 Financing the airline

Airlines have an enormous appetite for money, even when they are profitable. There seems to be a constant requirement to invest in new equipment, including aircraft, to upgrade and refurbish the customer's experience with such things as improved seats and easier processes, and to expand the airline's network. Airline profitability fluctuates and this together with the demand for funds means that the airline's management needs to watch the airline's current and forecast funding position all the time, not just in hard times. To do this, management should regularly look at the airline's current and forecast balance sheet together with short-term, medium-term and long-term forecasts of fund movements, surpluses and deficits. This information will be compared with the limits included in the financial part of the airline's strategy to ensure that the financial goals can be met. This is not an easy task against a background of demands for funding.

There are many ways of organising the finance for an airline's business. Banks and finance institutions frequently develop new ways to provide finance which will suit airlines in general or the circumstances of an individual airline. If this book tried to describe every method and source of finance, it would be out of date on the day it was published.

A simpler and more helpful approach is to trace and describe the financing of a fictitious airline from its foundation to maturity. Even this approach has its problems because in the real world the founders of an airline will have different plans on how to launch its operation. The two extremes are, to start in a small way with perhaps two or three aircraft operating on a small number of routes and, at the other end of the scale, to buy or lease a larger number of aircraft and rapidly operate a large number of routes. There are options between the extremes. Nevertheless, this chapter will assume the airline starts on a small scale and develops through stages to a larger, mature airline.

Risk

Before starting to discuss the types of finance available it is necessary to raise the issue of the basic risk involved in the financing plan. This is not a discussion of the risks involved in particular types of financing, but the fundamental decision of whether to use equity or loan finance.

If an airline's financial needs are entirely met by money from investors in exchange for shares, all the risks of failure and the benefits of success will rest with the investors. The investors pass money to the company and become entitled to the result arising from the operation of the business. If the result is a profit the investors are entitled to it, and if the result is a loss the investors bear the loss.

Taking finance from non-shareholders (e.g., banks), changes the shareholders' risk. The bank puts money into the company but does not share in the result of its operations. To compensate for this, the company is required to regularly pay the bank interest on the amount advanced to the company and, generally, to agree that if the company cannot repay the amount advanced when required, the company will sell some of its assets sufficient to repay the advance.

This agreement with the bank changes the shareholders' basic position. The interest payable to the bank will be paid from the operating result and, in doing so, will reduce the amount available for the shareholders. In addition, if the company's operation is not successful and it cannot repay the bank, assets will be sold. This probably reduces the potential future operating result and also reduces the amount that can be repaid to shareholders if the company is so unsuccessful that it ceases trading and decides to sell its assets.

This simple explanation illustrates how taking finance from non-shareholders changes the shareholders' risk. It also suggests the question 'why take finance from non-shareholders'? The answer is that if the extra finance enables the company to increase its operating result by more than the cost of the extra interest payable, the shareholders become entitled to the increased surplus.

It should be acknowledged that taking financing from third parties changes the shareholders' basic risk and increases the airline's risk because of the need to pay interest regardless of whether there is a surplus from its operations.

Funding an airline – before operations

When an individual or group of individuals is considering whether it is worthwhile forming an airline, a great deal of investigation is needed before a business plan can be developed. The costs of the investigation will either be funded by the individual or the group informally, by simply writing cheques for expenses as they are incurred, or by forming a company and paying money to the company in return for receiving shares and the company paying the costs as incurred. The contributors become shareholders and part of the equity referred to in Chapter 2.

Once the preliminary investigation is complete and if the founders conclude that it would be beneficial to form an airline, a detailed business plan will be prepared. This plan will cover the pre-operating period, and perhaps the first ten years of operations, for example, will include,

- The business philosophy, for example whether a low cost or full-service airline; whether to operate a cargo division.
- The airline's basic strategy, listing how it will differentiate itself from the competition; it may be too early to develop a full strategy for all areas of the business (e.g., finance, customer service), but this is the chance to consider producing the detail.
- Preferred routes and destinations, including present and future traffic rights and airport slots available.
- Aircraft needed, possible types, cost and dates needed.
- Other assets needed, offices, hangars, stocks together with the cost and dates needed.
- The amount, likely sources and timing of finance to acquire the assets.
- Staff recruitment.
- Risks including meeting aviation regulations, recruiting specialised staff and managing fuel costs.
- Forecast profits, cash-flow, financial position and dividends to shareholders.
- The work to be completed before the airline can start operations and the expected cost and timing for this stage.

The items included in the business plan must give a clear idea of the planned future of the airline, the potential and risks. At this stage it will be clear whether it is planned for the airline to start with a small- or large-scale operation. It will be used to raise funds from existing or new shareholders to get the embryonic airline to the point where it can start to build its operation. The new investors may be individuals or groups of individuals or companies known as 'venture capitalists', who invest money and receive shares in the company. Share certificates may exist as pieces of paper or in an electronic form.

When the airline has secured the initial funding, it can start to recruit staff to do the pre-operating development work, secure offices and start the process of converting the founder's ideas into an operating airline; this is a very detailed process. Much of the process involves developing systems and procedures to meet the regulations and laws controlling the airline business.

Funding an airline – operating

Once the pre-operating development work gets to the stage when it is clear that the airline will be able to meet all the regulatory requirements, the airline's management can decide how and when the airline can be launched. If the initial operations are to be on a small scale, the original business plan will probably be sufficient to secure more funds from shareholders or support a request for finance from financers. If the operation is to be on a larger scale, the shareholders will decide how much of the finance needed will be raised from shareholders and how much from financers and, at this point, the detail in the financial strategy will be helpful.

Additional funds may come from inviting shareholders to invest more, inviting new shareholders to invest, getting finance from financers or a combination of these, but in any event, an updated version of the business plan will be needed. The updated version will probably expand on the information in the original business plan by adding such things as the funding needed, the timing for advertising campaigns, and more accurate figures for the amount and timing to acquire assets such as aircraft, property, staff and different types of stocks.

At this stage, borrowing from third parties will be more difficult than for an established airline because the airline is not yet operating. Therefore, it cannot demonstrate that it will generate sufficient cash-flow to pay the interest on the debt or repay the debt itself.

Funding options

There are usually a variety of financing options available to an airline at any one time. All have their advantages and disadvantages. It can be that an option is more attractive or more possible at different stages in an airline's development. During the initial period, taking aircraft on operating leases and renting premises may be the most attractive course, but when the airline eventually has a financial history it may prefer to issue shares to the public. It is important that the airline's Finance Department stay in touch with the options currently available and understand their all-in costs, as well as their pros and cons. The funding options available change over time, as do the sources of funding.

When investigating current financing options, the most important items the Finance Department must bear in mind are the airline's financial strategy, the section on all-in cost, the security required and any existing financial relationships.

Short-term funding

Lease options

One of the options seriously considered by airlines when first starting is the 'operating lease'. This is an agreement where the airline has the right to use aircraft or other assets (e.g., vehicles), without acquiring ownership. The attraction of this for a new airline is because the initial outlay is less than buying an asset, but it is not unusual for an established airline to have some aircraft operating on leases in order to have the flexibility to release aircraft as part of the airline's fleet development plan.

At one stage, some airlines considered that one of the attractions of an operating lease was that it was not treated as a liability and was not recorded in the company's balance sheet, but was described in a separate note to the statutory accounts. From 2019, operating leases must be shown in the balance sheet amongst the assets and liabilities. This change is unlikely to change the business considerations on whether or not to use an operating lease, but it does change the presentation of the transaction in the company's accounts.

Operating leases can be used for many different types of assets ranging from aircraft to cars, and this option is always worth considering because it relieves the airline from the need to buy the asset.

It is frequently assumed that an operating lease is very flexible and the asset can be returned to the owner at any time. But this is not the case as most operating leases are for a minimum period, often three to five years. Any attempt by the airline to cancel the lease before the primary period has expired will involve penalties. The parties to an operating lease are the 'lessor', who is the owner, and the 'lessee' who is the user of the asset, i.e., the airline.

For aircraft, there are three types of operating lease. Most frequently the lease is a 'dry lease', which is just for the asset. A 'damp lease' consists of the aircraft and technical crew only. A 'wet lease' is for an aircraft and its technical and cabin crew. If the lease is for a very short period, the lease is often called a 'charter', hence there are wet charters or dry charters.

The advantages and disadvantages of an operating lease and the terms of the lease should be carefully evaluated. Not all operating leases are the same and what is seen as an advantage initially may become a disadvantage later, but in general the factors to be reviewed are as follows:

- Initially, the overriding advantage is that the purchase costs of the asset do not have to be financed by the airline.
- The asset can be returned to the lessor on specific dates. In future, this requirement may not be seen as attractive, because once the aircraft is integrated into a fleet and its maintenance programme, removing it may cause disruption to the business unless there is careful planning. Lease terms tend to be for a minimum period of three years with options to extend.
- The specification of the leased aircraft may not be the best for the airline. Lessors tend to buy aircraft and engines types which have a wide demand to maximise the chances of its aircraft being leased. If the aircraft's specifications are not optimal for the airline's route structure, the result may be higher route costs or maintenance costs, as well as an increased level of aircraft and engine spares.
- There are a number of payments involved in an operating lease, some of which must be paid in advance. Usually there is an initial deposit of, for example, three months rental, which is made in advance of delivery of the aircraft. The monthly rental is usually required to be paid in advance. In addition, there will be a requirement that the lessee pay amounts regularly to the lessor to create a reserve to cover maintenance costs. These funds can be released to the airline to meet the cost of repairs

and maintenance. Most operating leases require the lessee to pay an extra amount if the aircraft works more than an agreed number of hours or cycles in a month.

- When the aircraft is returned to the lessor essentially it should be in the same condition as when it was delivered to the lessee except for normal 'wear and tear', which is deterioration to be expected from normal use. Agreeing the return condition at the end of a lease can be difficult to negotiate and the cost of putting the aircraft into an agreed return condition at the end of the lease may be significant.

Although it varies with money market conditions, generally the all-in financial cost of an operating lease is more than owning the asset and arranging finance, but to offset this the airline does not have to finance the initial cost of the aircraft and it only has a commitment for less than the useful life of the asset.

Support from banks

Even at the early stage in the new airline's life it will be necessary to secure some form of short-term debt to support its operations. The receipt of income and the payment of expenses rarely match, and during the course of a month there are likely to be surpluses and deficits of funds in the airline's bank accounts. A common way to cover this problem is to arrange an 'overdraft facility'. This is the ability to temporarily pay from a bank account more than the balance in the account; it is a short-term loan by another name. The facility is usually not secured by a specific asset or assets but will have a maximum limit that can be drawn, and a higher rate of interest than on a loan will be charged because the bank does not know when the account will be over-drawn nor when the overdraft will be repaid.

There may be times when money in excess of the overdraft limit is needed to cover very short-term deficits (e.g., the costs of an emergency diversion of an aircraft to an airport it does not usually serve, or the time between buying an asset and arranging long-term financing for it). To cover these types of circumstances it is helpful to have a 'standby line of credit'. This is an agreement with a bank that they will advance a maximum amount of money for a short period on pre-agreed terms when requested by the airline. Usually a fee is paid when the line of credit is established, and interest will be paid on any amount advanced as and when an advance is made. It is usual for this type of facility to be available for one year, but it can be renewed each year.

As the airline develops it should become easier for it to raise any additional finance it needs because it will have a record of annual operating results. Potential investors and lenders will be able to see and examine the airline's financial position and form an opinion on the commercial and financial strengths of its management.

Longer-term finance options

An established airline will have a range of choices in ways to finance the business:

- Increasing the airline's equity by issuing additional shares either privately, usually called a 'placement', or to the general public, which is a 'public issue'.
- Issuing long-term debt either as a loan or as a 'bond'.
- Arranging 'finance leases'. This an agreement by which a third party owns an asset, often an aircraft, and leases it to the airline with the undertaking that the lease payments will repay the full cost.

The choice will depend on current market conditions as well as the effect on the airline's financial strategy.

Issuing shares

Often if the sum of money required is large, for example, to significantly increase the size of the aircraft fleet or build an office block or hangars, the preference is to issue more equity or to arrange a combination of equity and loans. It costs money to issue more shares, particularly to the public. The airline must give potential investors a clear picture of its business, its history, its current position and an assessment of the airline's prospects. This is done in a document called a 'prospectus'. The contents of a prospectus are set down in guidelines issued by the stock exchange on which the shares will be listed. The business contents will be similar to the information in the initial business plan, but in more detail, and in addition will include,

- experts' opinions of the value of the airline's assets.
- an audit opinion on the accuracy of the airline's financial history.
- commentary on the airline's financial history.
- a description of the airline industry in general and in the airline's particular region.
- an assessment of the future of the airline and the risks involved.
- opinions on other relevant matters, for example, any outstanding litigation or tax disputes.
- a description of how the money raised will be spent.

A public issue will involve a number of financial institutions, lawyers and auditors performing different advisory roles. In total these costs involved can be significant; perhaps 5% of the amount raised.

There is more flexibility in what needs to be disclosed if the share issue is a private placement, but the potential investors will certainly need similar information.

Equity can be thought of as a form of permanent debt. Just as the cost of a loan is interest, the cost of equity is 'dividends', which are the part of the profits earned and actually paid out to shareholders. The main difference is that loan interest has to be paid regardless of the airline's results, whereas dividends rely on there being profits.

Bank debt

A frequently used source of debt is a fixed-term commercial bank loan, repaid in agreed regular instalments, often every six months, during the loan period. This is called a 'self-liquidating loan'. The three main variations of the standard self-liquidating equal repayment loan are:

- 'Equal repayment loan', where the principal is repaid in regular instalments.
- 'Bullet loan', where the loan is repaid in full by one payment at the end of the loan term.
- 'Balloon loan', which is a combination of the two above. The repayments of the loan amount during the period of the loan do not fully repay the loan and a large final repayment is due at the end of the loan period. The balloon payment can either be repaid at the end of the loan term or the airline may take a new loan to repay it.

Bond

A 'bond' is a certificate evidencing a loan from a lender to the borrower (i.e., the airline). In the certificate, the borrower agrees to pay the holder a series of interest payments on a fixed date or a date(s) in the future, and to repay the amount originally lent.

There are two basic types of bond:

- Secured by the total assets of the airline, not any particular asset or aircraft.
- Secured against individual assets.

In most cases, a bond can be sold by the lender to a third party at any time. This gives lenders some flexibility on whether or not they hold the bond for its life, and the airline can cancel the debt by buying it back from the lender.

Finance lease

A finance lease differs from an operating lease because one of the underlying assumptions is that the cost of the asset leased will be repaid by lease payments. Hence, finance leases are usually long-term (e.g., for an aircraft it might be 12 years or more). The lessor will probably borrow money to finance part of the cost of the assets and the lease fees will take repayment of this loan into account, but the lessor remains responsible for the loan repayments. With this lease, the lessee has all the rights and obligations of ownership while it continues to pay the lease fees and complies with the requirements of the lease agreement. The advantage to the airline of a finance lease is that its cash-flow is more evenly spread over the life of the lease. Ideally, both the lessee and lessor benefit because they have different cash-flows arising from the ownership of the asset.

There is a form of lease which is often used to raise money against the security of an asset the airline owns (i.e., is not being used as security for any loan, such as an aircraft or building). It is known as a 'sale and lease back'. Under this agreement, a third party agrees to buy the asset at market value and lease it back to the airline. The agreement usually gives the new owner of the assets the right to sell it at the end of the lease period.

Finance sources

The main sources of finance for airlines are banks or large financial institutions such as insurance companies. A bank may act alone or with a group of other banks with one bank representing the group. This is called a 'syndicate' and the loan called a 'syndicated loan'. Generally, other financial institutions work alone.

Reserve fund and surpluses

In Chapter 2 it was mentioned that the airline's financial strategy would probably require that a level of funds be held in reserve to meet cash shortages in difficult times, such as in emergencies or to ensure the airline remained solvent. In addition to the reserve funds it is likely that an airline will have temporary cash surpluses. The surpluses can arise for many reasons, such as:

- Payments made in advance by passengers, whether they fly or not.
- Fluctuations in the airline's revenue streams during the year.
- Funds being accumulated to meet loan repayments and interest.
- Funds being accumulated to buy assets.
- Profits from airline operations.

The reserve fund and temporary surplus funds should be invested to earn interest for the airline, but the investments need to be as near as is possible risk-free and easily sold so the funds can be used by the business when needed. Although the investment requirements of the reserve funds and the temporary cash surpluses are the same, it is likely they will be invested separately and not combined. Reserve funds exist to cover emergencies. These do not happen frequently, hence the sort of investment that may be used is short-term government bonds, which are virtually risk-free and can be sold quickly, but do not pay the highest rate of interest. In this context, 'short-term' prob-ably means buying bonds which 'mature' (i.e., are repaid in full) after three months, and some which mature in six months. Many bonds issued by the government and by companies are 'rated'. Rating is a process by which an independent commercial rating agency (e.g., Standard & Poor's), which is actually a company but usually called an agency, assesses the risk of the issuer of the bond failing to make the required regular payments of interest and/or principal. This rating can be very helpful to investors when deciding which bonds to buy. Generally, the lower-rated bonds will pay a higher rate of inter-est to the holder, to some extent compensating for the increased risk compared to a highly rated bond.

The cash surpluses may be required day-by-day or at certain times to meet known payment dates. This requirement means that most cash surpluses will be kept on short-term deposit with banks or financial institutions. Where amounts are expected to be paid out on specific dates, the funds will be invested in a bond or deposit which matures on or near the required date(s). Maximising the return on these funds while minimising the risk of loss is the job of the Treasury function within the Finance Department. There is a potential conflict in having reserve funds to meet emergencies and investing the funds in profit-producing assets. This conflict needs to be resolved by the airline's manage-ment, weighing the benefits of a reputation for good financial management and financial risk management against the potential profits. It is not a problem that can be resolved purely with numbers. Reputation is a valuable commodity, but it is not possible to put a value on it.

Financial position

The airline's current and forecast financial position should be reported regu-larly by the Finance Department to the Board of Directors via the Finance Committee. The format of the report varies airline by airline, but it will be based on the airline's balance sheet, albeit in more detail and with a com-mentary. An illustration of a report for a fictitious airline is presented in Table 3.1.

A 'balance sheet' is a statement of the company's assets and liabilities as at a date and one of the statutory reports required to be prepared periodically by the company as part of its statutory reporting. When reporting to the directors with a management report, the report will show the current finan-cial position probably compared to the one previously forecast, together with a forecast of the position at the date the next report to shareholders is due, and for the following five or so reporting periods. The report should also show the current and forecast position compared to the financial strat-egy. The forecast prepared by the Finance Department will be based on the information received from operating departments for such major transac-tions as aircraft deliveries and substantial investments. Clearly, the further into the future the forecasts reach, the less likely they are to be accurate, but there should be sufficient confidence in the forecasts for directors to be forewarned of any potential problems which might arise. The report on the financial position will also contain information on 'working capital', which covers assets that can be readily realised in cash (e.g., debtors), these are also called 'floating assets' or 'circulating assets' and this will be discussed in Chapter 5.

Table 3.1 Extract from the finance report

	Last month		Financial year end			
	Forecast	Actual	2016	2017	2018	
Issued capital	38.4	38.4	38.4	48.4	48.4	Planned share issue in 06.2017
Retained profits	17.5	17.1	19.1	20.8	22.3	
Total equity	55.9	55.5	57.5	69.2	70.7	
Bonds						
Long-term	23.4	23.4	23.4	23.4	16.4	
Current	2.0	2.0	-0-	-0-	7.0	2016 bond repayment due in 07.2018
Leases						
Long-term	22.8	23.4	19.6	35.3	30.0	New aircraft delivered in 09.2017
Current	4.0	4.6	3.8	4.1	5.3	
Bank loans						
Long-term	9.2	9.2	7.8	6.4	5.0	
Current	2.8	2.8	1.4	1.4	1.4	
Total equity	55.9	55.5	57.5	69.2	70.7	
Total debt	64.2	65.4	56.0	70.6	65.1	
Equity %	46.5	45.9	50.7	49.5	52.1	
Debt %	53.5	54.1	49.3	50.5	47.9	
Equity 1: debt	1.15	1.18	0.97	1.02	0.92	

All figures are USD millions.

4   Financial risk management

Life is full of risks and occurrences which may result in a monetary loss or a bad result. Risk seems to be rather like gravity, something which people are aware of and adjust to despite it being invisible. People understand most of the risks in everyday life, and they take some action which makes them comfortable with its possible effects.

In general, when people deal with their business matters they also identify risks, decide on how likely they are going to happen and consider the size and severity of any possible losses, and then choose to manage them in a way that is appropriate for them. For example, everybody has identified the risk that their house or flat may be burgled and personal valuable items stolen. After thinking of the potential size of the loss, the chances of being burgled, the costs of ways to prevent the loss and the additional risks coming from their decision, they will decide what to do. There are a surprising number of options:

- Do nothing and take a chance they will not be burgled and, if they are, decide to replace the items out of their savings. In business this is called 'self-insurance'.
- Take out an insurance policy, whereby paying an insurance premium someone else takes the risk and will pay for any loss.
- Install an alarm system.
- Install a safe for the valuables and a system of security locks.
- Some combination of the above.

Apart from the first option, each of these methods of risk management involves taking on at least one new risk. For example, the insurance company may not pay every claim, or a power cut may immobilise the alarm system, or someone may forget to use the security locks or the lock of the safe is opened by an expert. Everyone understands these real-life risks and deals with them to the best of their knowledge and ability. Once a person makes the decision, they understand the risks and are comfortable with the solution and the risk they are taking.

The situation and process for dealing with risk are the same in business. Risks need to be identified, understood, evaluated and then managed in a way that results in the company being comfortable with the risks it has taken. There are many things that can happen in an airline, of whatever size, which could produce a monetary loss and/or affect the airline's reputation (for example, the adverse effects of major weather disruptions on an airline's schedule). The company has a duty to its staff and investors to manage these risks to the best of its ability. The company operates through its staff; therefore, the management of risk involves everybody to some degree. It is virtually impossible to eliminate risk, and so the options for managing it are essentially to:

- Monitor the risk and have sufficient resources available to minimise the bad consequences.
- Make business decisions which reduce the effect of an adverse occurrence.
- Look for ways to change the risk into a more acceptable form.

There are many risks which an airline must deal with and these include financial risks. It is a big subject and risk management is, or should be, a matter for the airline's Board of Directors to monitor. This chapter discusses the basic financial risks an airline is likely to have and the methods that can be used to manage them. It must be remembered that doing nothing is always an option in financial or other risk management, provided it is a conscious decision after identifying and evaluating the risk.

The major financial risks for an operating airline relate to commodities (e.g., primarily fuel) and money (e.g., mostly cash-flow and borrowings). The management of each risk usually requires the co-operation of a number of departments, both to provide information and to assist in the management process. The airline's Treasury will be significantly involved in risk management, sometimes supporting other departments and sometimes taking the lead, depending on the nature of the financial risk. There is a market where derivatives dealing with commodity risks, exchange rates and other financial risks are traded, and the staff in the airline's Treasury section will have the experience and expertise to trade in the market.

The Treasury section manages the operating financial risks of commodity prices and finance in a similar way that risk is managed in other parts of the airline, by organising its transactions to ensure the risks are acceptable. In addition, the Treasury may complete money market transactions through contracts, including derivatives, which makes the risk acceptable. To illustrate the techniques used by the Treasury section, it is helpful to look at the ways and techniques for managing three important financial risks: fuel price, interest rates and foreign currency exchange (forex) rates.

Managing commodity price risk

Airlines consume a number of commodities (e.g., wheat and beef for inflight meals, cotton for uniforms), but jet fuel is usually the largest single cost. It is useful to consider what the objectives of trying to manage the cost of commodities are. The objectives are usually laid down by the airline's directors and are a guide for the whole airline. The most frequent priorities are:

- First, to make costs as certain and predictable as possible.
- Then, to mitigate the effect of a sudden increase in the price or usage.
- Lastly, to benefit from any price or usage reductions as quickly as is possible given the earlier, higher priorities.

When discussing price risk management in this book, it is assumed that these are the airline's objectives.

The risks of changes in the prices of commodities, apart from fuel, are generally managed through the negotiation of fixed prices for the finished product because the airline does not use sufficient amounts of the non-fuel commodities to warrant the effort of using other approaches. Negotiating fixed prices for meals, uniforms, etc., effectively passes the commodity price risk back to the manufacturer. If the price risk cannot be passed to the manufacturer it can also be managed by substitution. For example, for inflight meals, if the price of beef rises unacceptably and the increase is reflected in the cost of the meal, it may be possible to substitute another meat for beef or change the dish altogether. Similarly, other materials may be used for uniforms.

Everyone who negotiates and agrees a contract has to bear in mind the need to try to manage all the risks of the transaction, including the financial risks. Before becoming involved in negotiating supply contracts, a manager should be educated in how to identify all of the risks involved, including price risk. After a manager has identified the risks in a transaction and developed ways to mitigate them through negotiation, there may be some risks that cannot be negotiated away (e.g., risk of increased cost due to changes in exchange rates or commodity prices). The airline's Treasury, together with operating management, can help manage these price risks.

Fuel price risk

Usually, an airline's main focus is on the usage and price of jet fuel, as fuel is a significant cost to an airline. A good current guide is about 30% of operating costs, but naturally this percentage changes as prices change. The fuel price risk arises because the price of fuel is usually based on an agreed margin over an index price (e.g., 'Jet Fuel Price Index'), and the index is based on the current market price for fuel, which changes day by day. The changes can be significant, and fuel is so important that most newspapers report the price each day. Fuel price risk is only one part of managing the risks associated with fuel; others include maintaining quality and ensuring the airline has a consistently reliable source in every airport it uses, including diversion airports. A review of the way fuel price risk can be managed illustrates how departments work together to achieve the objective.

Fuel price risk breaks down into two parts: price and financial exchange rate. Aviation fuel is usually sold in United States Dollars, often written 'USD', so if this is not the currency the airline uses and earns, there is a risk its fuel cost could increase purely because of a change in the rate of exchange between USD and the airline's reporting currency. Dealing with the currency risk is discussed later in this chapter.

Managing fuel price risk usually involves the airline's fuel purchasing section, the operations section and the Treasury section. The fuel purchasing and operations section will work together to establish the extent to which it is possible to reduce the airline's fuel cost by 'tankering'. This is to uplift more fuel than is needed at airports where the fuel is cheap in order to reduce the fuel uplifted at an airport where the fuel is more expensive. To do this effectively involves careful planning of aircraft and detailed calculations, but can produce savings.

When negotiating each fuel supply contract, the fuel purchasing section will attempt to minimise the margin over the index price to minimise the total cost. In addition, the negotiator may attempt to agree some fixed-price contracts and some where the price is averaged in an attempt to minimise the effect of price increases. Using fixed or averaged prices means the airline will not immediately benefit from any fall in prices.

The Treasury section's contribution to price risk management is to attempt to offset the effect of increases in the Jet Fuel Price Index by using 'derivatives', which are financial contracts that are traded on financial markets. Derivatives are always settled by one party paying money to the other; it is not necessary to handle physical fuel.

The Finance Department's Treasury section contributes to the management of fuel price risk by using derivatives. It is important to note that, generally, derivatives are used solely for risk management and not to make profits. The Treasury's main tools in managing fuel price risk or any other price risks are:

- A 'Cap', which is an agreement with a bank or other financial institution by which the bank guarantees a maximum fuel index price on a fixed date or dates for a defined volume of fuel. For this, the bank will charge a fee known as a 'premium'. If the actual fuel index price is more than the agreed level, the bank pays the difference to the airline. If the actual index price is less than the agreed level the bank does not pay anything but keeps the premium. This is like an insurance policy.
- A 'Collar', which is similar to a cap, but instead of the airline paying a non-refundable premium to the bank, the airline agrees to pay an amount to the bank if the index price falls below a certain level. Thus, there is an agreed upper level and an agreed lower level.

There are other forms of derivative and they are variations of the basic cap and collar. These standard forms of derivative are usually called 'plain vanilla'. There are derivatives which can be tailor-made for special circumstances; for example, linking fuel prices and interest rates, but these are unusual and are generally referred to as 'exotic'.

There are a couple of other ways an airline can try to minimise its cost of fuel, such as making a 'fuel surcharge' on all passengers and buying fuel-efficient aircraft. It is not always possible to introduce a fuel surcharge and, as soon as it is introduced, there will be commercial pressure to withdraw it, so while this can help mitigate the effect of rapid increases in fuel costs it cannot be relied upon. Many airlines are buying new aircraft models because they are fuel efficient. The benefits will accrue over time, but changing the composition of a fleet can take a long time.

Foreign exchange risk

The exchange rates between currencies can change significantly. A change of 10% or more over a few years is not unusual. For example, at the beginning of 2014 the exchange rate for USD to Japanese Yen, 'JPY' was about USD1 = JPY105. In 2015, this changed to approximately USD1 = JPY120, but by 2018 it was around USD1 = JPY 110. This sort of change can have a significant effect on an airline's income or costs.

Virtually all airlines have an exposure to changes in foreign currency exchange rates (forex), with the possible exception being those which operate only in the United States of America, but even these can face foreign exchange risks if they import products which are not priced in USD from outside the U.S. The price of aircraft and their spares are generally quoted in USD, so even domestic airlines in non-USD countries have a forex risk. An international airline may have cash-flows, income and costs in many different currencies.

The effects of changes in currency exchange rates are not always bad. If a country's currency becomes stronger, this may encourage more of their citizens to travel internationally for holidays. This will be good news for international airlines flying to and from the country, but it may be bad news for the country's domestic airlines.

The first stage of deciding whether to take any action to manage the foreign currency risks is to decide which 'currency exposures' i.e., the potential of a loss from changes in foreign currency exchange rate, need to be managed. The options are all, some or none. The decision on which approach to take will probably be made by the Board of Directors as part of the process of deciding what risks the airline is comfortable to have. There are arguments supporting and opposing each of the three options, and there is no one right answer for every company. Whichever option is taken needs to be kept under review because economic and political circumstances change, and these changes may also change the board's view of what is an acceptable risk.

Most international airlines will have surpluses in foreign currencies because a large part of their costs will be paid in the airline's home currency, e.g., head office costs including staff and flying staff costs. Unless they have significant receipts in USD, international airlines are also likely to be short of USD.

There are two ways to mitigate the effects of forex risk on income;

- Arrange a 'natural hedge', which essentially is to match receipts and payments in the same currency.
- Have a contract with a financial institution, usually a bank, to sell them surplus foreign currencies when they are remitted.

A natural hedge uses the foreign currency received to pay the local expenses and only remits to the head office any amounts which are surplus. The options are:

- Arranging to pay all expenses incurred in each country with the income received in that country.
- If there is still a substantial surplus of local receipts over local payments, it may be possible to cover the balance by borrowing the local currency to pay for aircraft or other assets. This approach will involve the airline paying interest at the local rate and this may or may not be attractive. The airline's Treasury will advise on the options.

Both of these are long-term arrangements, and this may be an unacceptable risk to the airline's management. The disadvantage of these types of arrangements is that, while the airline is protected from unfavourable changes in forex rates, it will not benefit from any favourable changes. The airline's management may take the approach that profits from favourable forex changes are windfall benefits and a low priority, as in the order of priorities discussed earlier.

If, after arranging the natural hedge, there is still a substantial surplus or management decide against the natural hedge option and there is a surplus, the option of making an arrangement with one or more bank or financial institution remains; the main options are,

- The airline can sell the expected surpluses of foreign currency in advance to a bank at an agreed exchange rate, called a 'forward sale'. There is a risk, though, that the surpluses will not arise at the time forecast, which would result in the airline paying compensation to the bank for the forward sale, but this may be an acceptable risk.
- Another approach is to make an 'option' contract with a bank, which gives the airline the right, but not the obligation, to require the bank to buy an agreed amount of the foreign currency at an agreed rate, at an agreed date in the future. For this, the bank will charge a fee known as a 'premium'. Whatever happens, the airline will pay the premium regardless of whether the option is used, sometimes described as 'exercised'. This is similar to a cap contract.

These risk management tools are frequently used to manage income receipts in various foreign currencies, but they can also be used where the airline has liabilities in foreign currencies (e.g., payments to aircraft manufacturers). In this case, when dealing with the bank, the forward sale becomes a forward purchase and the airline's option under the option contract is for the bank to pay the foreign currency amount to the airline.

Table 4.1 Extract from the cash forecast of a fictitious airline

FRA

Euro 000's

Month	FEB				MAR	APR	Next 3 months
Start balance	5						
Receipts	150	540	130	35	870	890	2,740
Payments							
Operations	35	40	35	42	160	170	500
Debt	-0-	-0-	700	-0-	50	47	800
Surplus	120	500	-0-	-0-	660	673	1,440
Deficit	-0-	-0-	605	7			

The example in Table 4.1 shows a problem which arises frequently in an international airline. This is the short-term cash forecast for a fictitious airline's operation in Frankfurt. It shows that the airline has decided to use a natural hedge, surplus receipts are used to pay all local payments and, in addition, the airline has arranged some form of financing in Euros. There is an overall surplus of Euro 108,000 in February, but there are expected surpluses in the first two weeks and deficits in the following two weeks. It falls to the airline's Treasury to resolve this problem; the options being,

- Locally invest, for the short-term, the surpluses of Euro 120,000 and Euro 500,000 as they arise and use them to fund the deficit later in the month.
- Remit surpluses to the airline's head office as they arise in the early part of the month and remit back sufficient funds to cover the forecast deficits. Depending on the current and forecast exchange rates, the Treasury may agree forward sales and purchases or options with a bank.
- Remit the surpluses arising in the first two weeks of February, arrange an overdraft to cover the deficits in the second two weeks of February and repay this from the surplus forecast for March.

The decision will depend on the current and forecast exchange rates, the interest likely to be earned from short-term Euro deposits and the rate of interest chargeable on over-drafts in Frankfurt. The calculation will also take into account the overall cash position at the airline's head office.

Interest rate risk

It is likely that an airline that has moved beyond the start-up stage will have medium- and long-term debts, and the interest cost can be significant. Interest rates change less frequently than foreign exchange rates, but this does not make it any easier to forecast the trend or timing of changes. Interest can be either 'fixed' for the length of the loan or 'floating' (i.e., changed at agreed intervals to the then current market rate). Having a fixed rate on each loan gives certainty about the amount of interest which will be paid, but may become an irritation if interest rate for loans fall; floating rates mean that the interest payable changes with the market and this can be seen as costly if rates rise. Clearly it is not possible to have the best of both worlds, hence in most cases an airline's Treasury will develop a model for what proportion of the airline's loans should pay interest at a fixed rate, and how much they should pay at a floating rate.

There will be times when the Treasury's model may not be appropriate, for example, when interest rates are at or near zero, and the weight of probability is that they will rise materially. There may also be times when the airline's management will wish to change the basis on which it pays interest, either from a fixed rate to a floating rate, or vice versa (as shown in Figures 4.1 and 4.2). If interest rates are expected to increase steadily and to remain at a new higher level for a significant period and the airline has a significant amount of debt on which it is paying interest of a floating rate, the airline's management may wish to change the interest payment basis from a floating to a fixed rate. To achieve this, the airline can arrange a 'basis swap' with a third party, not necessarily the bank which made the original loan. In this agreement, the airline agrees to pay the third party interest at a fixed rate, while the third party agrees to pay the airline interest based on the floating rate. The airline uses the money received from the third party to pay the lender, the net effect on the airline being that it is paying loan interest at the fixed rate. Figure 4.1 illustrates this.

Figure 4.1 Basis Swap from floating to fixed rate

If the airline is paying interest at a fixed rate and it is thought that rates will fall significantly in the near term and remain at a lower level, the Treasury may recommend swapping all, or some, fixed-rate loans to a floating rate. If this is done, the cash-flow will be as shown in Figure 4.2.

Figure 4.2 Basis Swap from fixed to floating rate

In both of these examples the airline's management is making the decision to change the basis of some loans based on opinion, perhaps just their own or perhaps reflecting the opinions of others.

In the same circumstances, other airline managements may be happy to adopt the Treasury's original model of the percentage of loans paying fixed and floating rates and leave it unchanged, despite actual or forecast changes in the loan market. Doing nothing is always a defensible option, provided that it is based on the careful consideration of the facts that are available.

The discussion on loan interest payable mirrors the discussion of interest which can be earned on various investments. Airlines, even small ones, have substantial cash-flows, and this presents the opportunity for the Treasury section to earn interest income on surplus funds. In addition, the accumulation of reserve funds as required by the financial strategy should be invested, even though the funds need to be available to meet emergency payments.

The same type of agreements with third parties are available to change the basis upon which the airline receives interest income from such investment as bonds. The considerations are the same, whether the airline should develop a model and not change it, or whether changes should be made depending on likely changes in the market.

It is often difficult to have a clear view of the trend in interest rates, so it is likely that for most of the time the airline's management will want to use the mixture of fixed and floating rates in the Treasury's model. Similarly, an international airline earning revenue in different currencies may decide to borrow in some of those foreign currencies, partly to achieve the natural hedge discussed earlier and partly to have access to different lenders. Again, the Treasury can develop a model of which currencies to borrow and in what proportion. If the airline does borrow in more than one currency and the airline's management would like to change the currency of a loan for a good reason, there is a financial contract called a 'currency swap' which will achieve this. The agreement will be between the airline and a bank or other financial institution to swap currencies on loan repayment dates at a rate of exchange which is fixed during the life of the swap agreement. In practice, this arrangement is similar to a series of forward sale agreements, but with an exchange rate which is fixed for the life of the agreement. As with interest rates, there is always the option to stay with the Treasury model and do nothing.

With all of these ways to potentially change the currency and interest rate basis, an airline's loan structure can become quite complicated and care is needed when reporting the position periodically to the airline's management and to investors in the airline's statutory reports. Table 4.2 presents an example of one way of reporting the loan analysis of a fictitious airline.

Table 4.2 Loan analysis of a fictitious airline

Currency borrowed	Interest rate		TOTAL
	Fixed	Floating	
GBP £ pounds	1.52	3.28	4.80
USD $ dollars	.25	1.3	1.55
JPY ¥ yen	1.07	-0-	1.07
Total	**2.84**	**4.58**	**7.42**
Repayable (years)			
Less than 1	-0-	.68	.68
1–2	.50	.06	.56
2–5	1.00	.92	1.92
More than 5	1.34	2.92	4.26
Total	**2.84**	**4.58**	**7.42**

Figures are reported in USD equivalents.

Derivative risk

The third-party agreements mentioned in this chapter (i.e., forward currency sales and purchases, and the basis and currency swaps) are all forms of derivatives. A derivative is an agreement related to the value of a transaction, but not necessarily between the same parties as the original transaction. Using derivatives to manage an airline's financial risks introduces some new risks to the airline. The main risk is:

- A 'counter party risk', which is the risk that the third party will not honour their side of the agreement. This risk is usually managed by setting 'trading limits', which involves having a maximum limit on the total value of outstanding transactions, whether assets such as deposits, or liabilities such as loans with every bank or financial institution. This is the same approach which banks use to manage their counter party risk, but assessing the creditworthiness of a bank or financial institution and allocating a trading limit to it can be very difficult. It is also useful to have a limit on the number of transactions which can be made with each institution each day or each week.

However, a significant risk is for a company to use a derivative where all the risks of the derivative are not completely understood by its Treasury team, Finance Committee and its Board of Directors. There have been examples of companies incurring significant losses because the risks involved in a derivative have not been fully understood and there has not been an attempt to manage them. Every person responsible for recommending, approving or trading derivatives should understand how each derivative works and the risks involved. There are courses and books on this topic.

Reporting requirements

One of the Finance Department's important functions is reporting the airline's current and forecast financial position to the airline's management and its directors. These reports are essential to help the management and directors guide the airline and to ensure the airline remains solvent. The cash forecast is a key document to remaining solvent, and all the airline's departments have a part to play in producing a realistic cash forecast. All the basic information used for the cash forecast comes from the operating departments. Generally, the finance report will be made monthly, but there may be good reasons for producing it, or part of it, more frequently. A major problem is to produce the finance report quickly yet accurately.

Speed and accuracy are often in opposition and it may be necessary for the board to give guidance on when the information is required (e.g., one day after the end of the month), and also what level of accuracy is acceptable. It is important that the figures are as current as possible because the report is read against the background of the financial markets, which often change quickly.

Designing the report is also difficult. It needs to cover the complete financial position of the airline but there is the danger of including too much detail. Key information may be difficult to see because it is buried by non-actionable data, and this is the well-known problem of 'information overload'.

Every airline will have its own format for its finance report because its circumstances and priorities will be unique. For example, a domestic airline will have a very simple section on currencies, but dealing with a large number of currencies may be a major area of concern to an international airline. The report may be produced in various formats depending on the needs and responsibilities of those receiving the report. It is likely that the members of the Finance Committee will require a more detailed report than the directors on the board, but given the importance of such items as the fuel cost to an airline, the directors may require full details of the current hedging position.

The essential contents of the report are:

- A summary of the current cash positions, together with a cash forecast for the next three, six or 12 months. All funds and all locations should be identified. There should be an explanation of the reasons for difference between the actual and forecast cash-flow.
- A similar section reporting the current and forecast position for foreign currency cash-flows together with comments on any significant changes from the latest forecast. It is likely that the Finance Committee, as part of the airline's general risk management, will have agreed limits on the percentage amount of foreign currencies that may be committed to debt servicing, loan interest and repayments and, if so, the actual and permitted percentages should be noted.
- A list of the outstanding financial agreements, derivatives and other agreements used in managing the airline's financial risks, together with any

plans to enter into new, or to terminate any current risk management, arrangements.

- Detail of the short-term investments (e.g., of surplus funds or the reserve fund), together with a note of the earnings from each investment compared to any agreed target rate.
- A list of the current and planned long-term financings of the airline, equity and all forms of borrowings. This was discussed in Chapter 3 and is shown in Table 3.1. An analysis of the currency and interest profile of a fictitious airline is shown in Table 4.2.
- A statement of the outstanding value of commitments made to banks and other third parties for all risk management agreements and loans. The statement should also include amounts owed to the airline by banks and third parties (e.g., bank balances, deposits and derivatives). It should be remembered that banks and financial institutions periodically cease trading, and in this case the airline might suffer a loss.

It is likely that the Finance Committee will have authorised the Treasury to complete transactions in certain circumstances (e.g., change the interest rate on the loan from XYZ Bank from floating rate to fixed rate when the fixed rate is 'X%' or better). These pending transactions should be listed.

Periodically, perhaps once or twice each year, the Finance Department will produce a long-term forecast, covering perhaps three or five years of profits and the airline's financial position, using information supplied by each department. The forecast will give management and directors an indication of what the airline could look like in the future and also indicate where problems may arise. It is the Board of Directors' responsibility to work on solving problems before they arise. The longer the period covered by the forecast, the less accurate the figures are likely to be, but operating departments need to do the best they can to look into the future. The better the quality of the input, the better the decisions will be, leading to a better outcome. Managing an airline successfully requires long-term planning, as well as efficient daily operations.

It takes time to make changes in an airline because many decisions take considerable time to implement (e.g., when and what type of aircraft to order). Similarly, changing the airline's financial structure cannot be done overnight. The long-term financial forecast is an educated guess. Hence it must not be given more credibility than it deserves. Nevertheless, it is the best guess the management can make at the time. There are various approaches to producing long-term forecasts. Some will be just one set of figures supported by the assumptions and the information provided by the airline's operating departments, while others will have a number of alternative forecasts (e.g., best, most likely and worst). The airline's management must judge the amount of detail required and the effort required to produce the information, and there is no point assuming that the forecast is anything other than just that: a forecast.

5 Managing the airline's financial position

Information on an airline's financial position is important to people both inside and outside the airline. Inside the airline, its management will want to monitor both the current and forecast financial position regularly to ensure the airline is still solvent and that the money invested in the business is being used efficiently. Outside the airline, current and potential investors, lenders and suppliers also want to see the airline's current position; it is unusual for outsiders to get information on the forecast finances except when the airline issues new shares. The investors, lenders and suppliers are interested in the same things as the airline's management: solvency and the efficient use of funds. It is often forgotten that suppliers frequently look at a company's financial history, particularly its debtors/creditors/liquid funds position, when deciding whether to do business with it.

The balance sheet

The airline's financial position is shown in a document often called a 'balance sheet' or a 'statement of financial position'. Essentially, the report shows what the airline owns and what it owes. The balance sheet is part of the company's statutory reports, but the format of the report to the airline's management may be different from the statutory report. The figures, however, will be the same. Regardless of format the function is the same; to provide information to help the readers assess the airline's financial strength and financial management.

The contents of the airline's balance sheet are strongly guided by professional reporting guidelines, which apply to all companies. Unfortunately, the format of the statutory balance sheet is not always inviting or easy to read and use by the average person, but there is little a company can do to change the format except to add additional information and explanation of the figures in notes to the financial statements. The balance sheet used by the airline's management will contain the same information as in the statutory reports, but there is a great deal of flexibility in the way the information is presented. The format of the management balance sheet is driven by what the airline's management want to see and use. It must be easy to read and use a layout which highlights important information. Like all management reports, the format of the management balance sheet will vary company by company, but will include the same information as the statutory balance sheet, albeit usually in more detail.

In practice, there are few differences between the balance sheet of a Low-Cost Carrier (LCC), and a Full-Service Airline (FSA). Generally, there are more items included in the balance sheet of a Full-Service Airline (for example, mileage programme and catering stocks), but these just reflect the difference in business models. Whichever business model is used, normally the three largest items in an airline's balance sheet are its long term assets (mostly the value of aircraft), borrowings to finance the assets and liquid funds. Like most businesses, the balance sheet of a start-up airline is usually simpler than that of a mature operation. As the business grows and the total figures become larger, balance sheets tend to be more difficult to read easily, and the detail in the notes to the balance sheet becomes increasingly important.

Accounting

At this stage it is useful to discuss 'accounting', which is the skill of correctly recording regularly and systematically in the airline's 'books of account' or records, all its business transactions that result in the transfer of money or money's worth, so the reader knows the true state of the business. In other words, the purpose of accounting is to record and communicate the true financial state of a business.

The current system of accounting is known as 'double entry', and is based on a very old system allegedly developed by monks. The approach assumes that every transaction always has two aspects, an increase or decrease in an asset, and an increase or decrease in a liability. An example of this being if you buy clothes with cash, your cash reduces, but your clothing asset increases. The advantages of the double-entry approach are as follows:

- There is a complete record of each transaction.
- The accounts contain full financial information; hence statements showing the achievements and the state of the financial company can be produced.
- The records cover all aspects of the business, making control easier and revealing trends.
- There is a check on the arithmetic accuracy of the clerical process because the same total value is used for both entries in the accounts.
- It is simple to understand.

The theory supporting accounting is that in every business transaction there is a transfer of value between two parties. This value may be in the form of money, goods or services. Usually, goods or services will be exchanged for money, but there are arrangements where goods or services are exchanged for other goods or services (e.g., airline tickets) without money changing hands; this is called a 'contra deal'. When the transaction takes place, it is assumed that the values exchanged are equal, hence the exchange is balanced so the accounts recording the transaction are also in balance. Accounting entries are made on the basis of an 'accrual', that is, they record the value of a transaction when value passes, not when payment is made (for example, when a service is performed, not when the invoice for the service is paid).

Underlying all accounting practices is the very basic idea of 'balance', which acknowledges that every transaction has at least two aspects and that the total of the value gained (an 'asset') is matched by a reduction in another asset and/or an increase in an amount owed (a 'liability'). Thus, it can be seen how the term 'balance sheet' arises. The examples in Tables 5.1a, b and c show how the basic accounting works in practice. In this example, a newly formed airline issues shares to investors in exchange for USD 10,000,000; the airline then buys an aircraft for USD 20,000,000, using a 10-year bank loan of USD 15,000,000 and USD 5,000,000 of its own funds. The balance sheets at the end of each stage are shown in Tables 5.1a (after issuing shares) and 5.1b (then after buying the aircraft).

Table 5.1a Balance sheet after issuing shares

Balance sheet as at 31 December 20XX

	USD		USD
Capital and reserves	10,000,000	Liquid funds	10,000,000
Total liabilities	10,000,000	Total assets	10,000,000

Table 5.1b Balance sheet after buying aircraft

Balance sheet as at 31 December 20XX

	USD		USD
Capital and reserves	10,000,000	Fixed assets	20,000,000
Bank loan	15,000,000	Liquid funds	5,000,000
Total liabilities	25,000,000	Total assets	25,000,000

It can be seen that at the end of each stage the accounts 'balance'.

There is one point of detail that is worth mentioning. Many people when reading a balance sheet for the first time wonder why the amount invested by shareholders is recorded as a liability. This is because the money is owed to the shareholders; if the airline stops operation and all its assets are sold and liabilities settled, the shareholders will be repaid from the surplus.

It is usual to present the balance sheet in a vertical format separating items between 'non-current assets and liabilities' (the items that are held for more than 12 months), and 'current assets and liabilities. This helps the reader to assess whether the company has sufficient funds in the short term. Frequently, the figures will be rounded-off and shown in millions only to help the reader understand the figures. The balance sheet in Table 5.1c presents the same information in a vertical format.

Table 5.1c Balance sheet in vertical format

Balance sheet as at 31 December 20XX

	USD million
Non-current assets	
Fixed assets	20
Long term liabilities	15
Net non-current assets	5
Current assets	
Liquid funds	5
Net assets	10
Capital and reserves	
Issued capital	10
Total equity and reserves	10

Once again, the figures balance and, in addition, the revised format shows,

- the airline's long-term assets exceed its long-term liabilities, which is good
- there is substantial money in the bank, which is also good.

From the figures in the balance sheet it is possible to calculate the airline's debt/equity ratio of 1:1.5 (USD 10 million/USD 15 million) and it shows that the airline is funded 40% by equity and 60% by debt. The airline's management will compare this to the figure in the airline's financial strategy.

Once the airline starts its operations, many more accounting entries will pass through its accounts and the balance sheet will become more complicated. Table 5.2 is an example of a fictitious airline's balance sheet as it would appear in the statutory reports.

The terms used in the balance sheet are as follows:

- 'Assets' are the property and possessions of the business (e.g., land and aircraft).
- 'Fixed assets', also called 'non-current assets' are the assets to be used for many years to generate income (e.g., aircraft).
- 'Intangible assets' are assets which cannot be seen (e.g., computer programmes, trademarks).
- 'Investments in associates' are the amount the airline has invested in other companies (e.g., a company that performs ground handling). The definition of 'associate' can vary between countries. The usual definition is a company where the airline owns a significant percentage of the company's

Table 5.2 Balance sheet of a fictitious airline

Balance sheet of Fictitious Airline Ltd.	
As at 31 December 20XX	
	USD *million*
Non-current assets and liabilities	
Fixed assets	98
Intangible assets	10
Investment in associates	23
Other long-term receivables and investments	6
Total	137
Long term liabilities	56
Other long-term payables	2
Deferred tax liabilities	10
Total	68
Net non-current assets	69
Current assets and liabilities	
Stocks	2
Trade and other receivables	11
Liquid funds	21
Total	34
Current portion of long-term debt	10
Trade and other payables	24
Unearned transportation revenue	12
Total	46
Net current liabilities	(12)
Net assets	57
Capital and reserves	
Issued capital	17
Reserves	40
Total equity	57

shares but does not own the majority, yet it can influence the company, often by appointing a Director to the company's Board of Directors.

- 'Other long-term receivables and investments' are all amounts receivable after the next 12 months, including investments in other companies the airline cannot influence.
- 'Long-term liabilities' are debts related to the purchase of assets which are due to be paid after the next 12 months.
- 'Other long-term payables' are any other amounts due for repayment after the next 12 months.
- 'Deferred tax liabilities' is the amount of a temporary difference between taxes that will become due in the future, and taxes payable today.

- 'Net non-current assets' are the difference between the long-term assets and the long-term liabilities.
- 'Current assets' are assets that can be readily converted into cash. Sometimes these are called 'floating' or 'circulating' assets.
- 'Current liabilities' are amounts due to be paid in the following 12 months.
- 'Stock' includes all types of items held for use in the company's business (e.g., engineering spares parts, inflight services items).
- 'Trade and other receivables' are all amounts due to be paid to the company within the next 12 months. If any debts are due to be paid after more than a year they will be reported in 'other long-term receivables and investments'.
- 'Liquid funds' includes cash and investments that can quickly be converted into cash.
- 'Current portion of long-term debt' is the amount of the long-term debt to be repaid within the next 12 months.
- 'Trade and other payables' is the amount due to suppliers for goods and services and includes approved invoices waiting to be paid and estimates for invoices not yet received.
- 'Unearned transportation revenue' is the amount received from passengers and shippers in advance.
- 'Net current liabilities' are the difference between current assets and liabilities; in this case, current liabilities exceed current assets, but it may be the other way around.
- 'Net assets' are the difference between net non-current assets and net current liabilities.
- 'Issued capital' is the amount received by the company for the shares it has issued. This is not the same as the current market value of the company's shares.
- 'Reserves' is the amount of profit which has not been paid to shareholders as dividends.
- 'Total equity' is the total amount the shareholders have invested in the company, and this total agrees with the amount shown as the company's net assets.

Notes attached to the balance sheet will give more detail about each item and the previous year's figures will be shown alongside each item as a comparison. It is possible that airlines will use other terms and may change the layout, but the balance sheet in Table 5.2 shows the basic layout.

Assessing and using the balance sheet

A balance sheet is not included in a company's statutory reports simply to complete the picture of the company, but to give the reader information on the amount of money being used in the business, where it came from and what it has been used for. This is important information for every reader. Before using the information in the balance sheet, it is useful to discuss the basis for the figures and how accurate they are.

In the vast majority of cases, the statutory financial reports are prepared on the basis that the company is still a 'going concern', that is, it is assumed that the company has sufficient resources of money and staff to continue operating in the foreseeable future and to meet its obligation when they are due. The main consequences of this assumption are,

- it is not necessary to adjust the value of assets in the balance sheet, e.g., aircraft, for any fall in their market value if the fall in value is expected to be short-term
- expenditure on projects, e.g., major new computer projects, can continue to be carried in the balance sheet and written off over more than one year.

The alternative to a going concern basis is a 'liquidation basis'. As the name says, this assumes the company will be or is being liquidated, i.e., is ceasing to trade and is selling assets and repaying obligations. If this basis is used, all assets are recorded at their expected sale value and all liabilities are stated at the full amount that is due, and this may include penalties for early repayment in the case of some loans. Additionally, an estimate of any expenses incurred solely relating to the liquidation will be included.

Readers of the statutory financial reports get an assurance of their accuracy from the opinion of the company's external auditors following their review of them. If the reports do not contain any materially incorrect valuations and are complete, the external auditors will confirm that the reports give a 'true and fair view' of the financial position of the company. This true and fair view means that, in the auditor's opinion, the financial reports do not contain any material misstatements and accurately represent the financial performance and position of the company. If there are any significant misstatements or omissions, the auditors will give a 'qualified opinion' or decline to give any opinion. Having a qualified opinion is, of course, an important issue for the company and one that should be resolved quickly.

Readers of the balance sheet need to bear in mind that it contains a number of estimates. The estimates are the best the company can make to produce an accurate result and they are reviewed by the company's external auditors; nevertheless, they remain estimates. The most important areas involving estimates are:

- The value of non-current assets (e.g., aircraft). The value in the balance sheet is the original cost less the depreciation already charged as a cost in the Profit and Loss Account. The amount charged as depreciation is the cost of each asset less its estimated value at the end of its estimated useful life to the company and this net amount is spread over the period of the asset's useful life to the company. It is difficult to estimate the sale or scrap value of an aircraft 15 or more years in advance. Similarly, fleet plans may change and aircraft may be kept for longer or shorter periods than originally planned. It becomes easier to estimate the final value and useful life of an asset as it approaches the end of its useful life to the airline.

- Some maintenance costs are incurred over long periods (e.g., an aircraft's C and D checks), each year an estimate of these costs has to be made and an amount set aside to pay them in the future. The amount set aside is included in the balance sheet.
- If the company has a dispute (e.g., with a supplier), which has not been resolved at the date of the balance sheet, an estimate of any amount the company may have to pay when the dispute has been resolved will be included in the balance sheet. If the amount cannot be estimated the notes to the accounts will mention the dispute.

The statutory balance sheet

The format in Table 5.2 is the one generally used for statutory reporting. The statutory balance sheet will be read mainly by the airline's shareholders, financial institutions and suppliers of goods and services. While the balance sheet figures show the airline's financial position, the figures alone do not immediately indicate how efficiently the airline's management is using the funds at its disposal and whether there is a trend, nor does it permit easy comparison to other airlines and industry averages. It is often difficult to 'use' total sums of money to help management run the airline, as the figures need to be analysed and related to the airline's operations. To assess the balance sheet figures and to compare them with the airline's current and future plan and make other comparisons, the figures in the balance sheet are often converted into ratios. There are some fairly standard ratios that can be used. These standard ratios permit comparison with an individual airline, industry averages and indeed companies in other industries. The main ratios are:

- 'Debt/equity ratio', which was discussed in Chapter 2. It should be noted, using Table 5.2, the total debt is long-term liabilities plus current portion of long-term debt (USD 56 million + USD 10 million = USD 66 million). Equity is the issued capital plus reserves (USD 57 million). The calculation is USD 66 million ÷ USD 57 million = 1.16 times. Another presentation is: debt 53.7%, equity 46.3%.
- 'Percentage return on capital employed'. 'Return' is another way of saying 'profit'. The calculation is:
  Profit / Equity (USD 57 million) + Borrowings (USD 66 million)
    Assuming the profit was USD 10 million, the calculation is USD 10 million ÷ USD 123 million = 0.081 or 8.1%
- Percentage return on equity. The calculation is:
  Profit / Equity
    The calculation is USD 10 million ÷ USD 57 million = 0.175 or 17.5%
- 'Times assets turned over', which is,

Revenue / Assets

Assuming the revenue was USD 475 million, the calculation is USD 475 million ÷ (USD 137 million + USD 34 million) = 2.78 times

- 'Quick ratio', which is sometimes called 'working capital ratio', and indicates whether a company has the resources to pay its short-term debts. 'Working capital' is the term for current assets and current liabilities.

Current assets / Current liabilities

The calculation is USD 34 million ÷ USD 46 million = 0.74

In summary, the figures are:

Debt/equity ratio	1.16 times
Percentage return on capital employed	8.1%
Percentage return on equity	17.5%
Times assets turned over	2.78 times
Quick ratio	0.74

It is difficult to objectively assess whether or not these ratios are acceptable. The opinion will depend on what the airline's management had planned, what other similarly financed airlines have achieved, and what shareholders expect. Each reader of the balance sheet in the airline's statutory report will, in addition to the standard ratios, apply their own measures, ratios and standards to assess what the airline's financial position is. An example is that the reader of Table 5.2 may conclude that the amount received in advance of travel and included in current liabilities should not be included in the quick ratio because it is matched with cash holding in current assets. In this case, the calculation becomes (USD 34 million − USD 12 million) ÷ (USD 46 million − USD 12 million) = 0.65.

Balance sheet for management

The version of the balance sheet used by the airline's management will have the same information as the statutory report, but it will probably have a different layout and show more detail for some items (e.g., stocks, debtors). The figures in the airline's management balance sheet and the ratios derived from them will be compared with the airline's financial strategy and its current and future plans, and this comparison will show whether the trends or ratios are acceptable to the management. Clearly the figures in the airline's strategy will become Key Performance Indicators (also referred to as 'KPIs'), which are the set of indicators by which the airline judges its success. These are discussed later in more detail in Chapter 7. This comparison probably does not give the management all the information it needs; the financial figures must be related to what has happened and what is planned to happen within the business.

There will probably be a list of other ratios which are used to examine the balance sheet. The list generally splits into two parts: permanent ratios designed to highlight the trend of items that must be controlled (e.g., debtors), and temporary ratios which deal with areas which are currently a problem.

An example of a permanent ratio might be the ratio of the amount invested in engineering stocks, which are part of current assets, compared to the value or the number of aircraft, which are non-current assets. This might indicate whether the amount invested in the stocks is worryingly high or low. Another example is the ratio between liquid funds and the amount owed to creditors; both of these items are in the list of current assets and liabilities. A low ratio may indicate the potential for a solvency problem.

A temporary problem may be that the amount invested in trade debtors, part of current assets, has been agreed as being too large and the action to reduce the amount may be measured against the airline's revenues and/or number of passengers; if the worry relates to non-passenger/shippers' debts, the amount for debtors may be related to ancillary revenue.

Controlling the amount invested in the business is of vital importance, as investments should be made to produce a return. If the amount invested in stocks, debtors and prepayments can be safely reduced, the amount released could be invested in assets which produce a return.

Frequently, the format of the management balance sheet is looked at as being in two sections; the first is the non-current asset section which deals with assets held for the long term (e.g., aircraft, land and investments in related companies). The second section is current assets, which are changing all the time, such as the value of stocks held which change every day.

With non-current assets, the management of the amount invested usually takes place when decisions are made as to whether and when to buy or sell a major asset. The decision is likely to include a decision on financing as well. Once a decision has been made, the effects will flow through into the non-current assets and long-term liability part of the balance sheet. This is not to say that the position of non-current assets and the related liabilities should not be reviewed regularly by the airline's management. These periodic reviews may produce a recommendation to make some long-term change, but the frequency of change is likely to be significantly less than for current assets and liabilities.

Managing working capital

Current assets

Decisions on buying or selling some non-current assets often influence the airline's working capital. Examples are as follows:

- A decision to buy or lease additional aircraft is likely to affect the level of engineering stocks and inflight service stocks held.
- The decision to sell an aircraft might reduce the level of engineering emergency spares needed, but this is not always the case as a minimum level may be needed regardless of fleet size.

- The decision to rent a new office may involve making a deposit and paying some rent in advance.

The likely effects on working capital should be included in the proposal for all decisions made on non-current assets. The airline's working capital uses the airline's funds, and this has a cost. Hence, the amount invested in this area needs to be controlled. Working capital, current assets and liabilities, are managed day-by-day, by both routine and exceptional decisions made by individual managers throughout the airline. Individual purchase orders are placed, which may include deposits or advance payments, invoices are paid, and passengers book and pay for tickets. All of these go on every day, and each one changes the amount of working capital in the airline. The amount of working capital must be closely monitored because it can change rapidly and/or the amount of tied-up funds may become unacceptable to management.

One part of managing working capital is to analyse all of the items in the working capital section of the management balance sheet into the amounts, which is the responsibility of a designated manager who is solely responsible for controlling the amount involved, and for understanding the factors which influence the total amount. Frequently, the manager responsible will be a department manager. The analysis of the airline's working capital into its component parts and the identification of the responsible manager, is designed to ensure that each part for working capital is the responsibility, and in specific cases, the undivided responsibility of a manager who can control the total monetary figure and influences the measure or ratio used to control it. It is also important to ensure that each item and part of working capital is clearly related to a specific operating activity. For example, stocks may be broken down into engineering stocks, inflight services stocks and marketing stocks. If the figures are large enough, these categories can be further broken down (e.g., engineering stocks into routine spares, emergency spares and general engineering parts). The objective is to have clear management responsibility for a part of the balance sheet.

Another part of managing working capital is to decide how each category in the working capital account should be measured and monitored. The monetary amount in each category should be measured against the activity in the airline to which it relates as an attempt to assess whether the amount invested is either too much or too little. An example is to measure the value of the various engineering stocks held for each aircraft type with the average value of spares used each month, or the number of block hours for each aircraft type, the flight cycles flown by each aircraft type, or by some other measure. The important principle is that the measure has to be accepted by managers as a valid measure of activity for the working capital item. Every heading in the management balance sheet will be treated in the same way (i.e., analysed into responsibilities with one or more measures agreed) and the figures will be reviewed regularly. If an unacceptable trend develops then action needs to be taken to correct the problem. It is likely that management will want to see the controlling ratios and measures for more than one year because problems tend to develop over time rather than suddenly, therefore an unfavourable trend may need to be identified before action is taken.

Where it is possible, there should be an agreed policy for dealing with items in current assets. The most obvious example is a policy on debtors (i.e., deciding) who should be allowed credit, how much and how the amount due should be controlled. This will be discussed in more detail in the next chapter. It may be possible to construct a workable policy for other working capital items.

The ratios produced from the actual figures in the balance sheet should be compared with a target for the ratios set in the airline's short-term plan, just as the actual measures for the non-current assets and liabilities will be compared with the financial strategy. Any deviation will be investigated, and corrective action taken. Once the action and the time required to complete the action have been agreed, it is possible to produce amended measures which can be used to monitor future figures.

It is useful to compare all the ratios and measures used to manage the balance sheet (i.e., the airline's financial position) with those in other airlines, particularly competing carriers. Unfortunately, this tends to be difficult to do because the other carriers' statutory reports frequently do not include the equivalent information nor provide sufficient detail to enable calculations to be made. Nevertheless, it is useful to gather what information can be gleaned from the statutory reports of other airlines to try and see whether they are financially more efficient in some key areas.

If the airline's operations are stable (i.e., there are not any changes in the composition of the fleet, the routes the aircraft fly, or the way the airline's operations are organised) the measures used for managing working capital do not need to be changed, but periodically they should be reviewed to ensure they are still the most appropriate. Most airlines instigate changes, sometimes significantly, for example:

- changes in the composition of the aircraft fleet to introduce more fuel-efficient aircraft
- reconfiguring cabin interiors to improve facilities for passengers
- opening offices in new destinations.

All of these involve short-term changes in the airline's working capital and this needs to be reflected in the measures used to assess the amount invested in working capital. Otherwise, the measures will not be valid comparisons and may mislead management. Although the changes may be short-term, that 'short term' may stretch over several years for such changes as a change in fleet composition. It is important that any change is justified as part of management's review process and advised to the managers responsible for the working capital items involved. The change should be formally approved by the committee responsible for approving changes. The likely financial implications of the change should be noted and if it is not possible to calculate an exact figure; there should be an indication of the magnitude of the change, e.g., 'X ratio is likely to increase by between 4 and 6 percentage points'. If the ratio or measure for an area of working capital needs to be revised, temporarily or permanently, the manager responsible will recommend to the CEO for their approval a revised measure for each of the years involved in the change. In some cases, a management committee may have the authority to approve changes; it all depends on how the airline is organised. The recommendation will be made after consulting with the Finance Department to establish the accounting treatment and possible timing for the change. Once agreed the amended measure can be used for monitoring future actual balance sheet values.

Current liabilities

The airline's current liabilities also have to be managed (see Table 5.2). The main areas are Trade Creditors and Unearned Transportation Income, both of which have to be managed. The current portion of long-term liabilities is managed when the detail of the original loan is negotiated. Both Trade Creditors and Unearned Transportation Income reduce the amount of capital employed in the airline so are not a cost, but still need to be managed.

The airline will have an agreed policy for dealing with creditors just as it has for dealing with debtors. This policy will include a maximum time by which creditors' invoices should be paid, and also the method by which any dispute on an unpaid invoice is to be resolved. The measure for creditors will usually be the numbers of days of average operating cost, excluding salaries. Dealing with creditors has an extra consideration for management as there is the risk that if creditors are handled badly, the airline's reputation will suffer.

The measure for Unearned Transportation Income will usually consist of days of average sales, the number of advanced bookings or the amount per passenger (this is more likely to be used by an LCC). On the surface it may seem very attractive to have a very high level of Unearned Transportation Income because it is interest-free money, but often when passengers book far in advance, it is in order to secure a lower fare.

The agreed ratios or measures for working capital are an important part of the airline's planning and forecasting process. There is nothing an airline can do about what has happened in the past except to analyse it, explain it and learn from it. The lessons learnt will be reflected in the decisions that will influence the airline's future. Everyone in the airline is involved in shaping the airline's future, whether that future is tomorrow or in several years' time.

So far, the discussion of managing an airline's balance sheet has been in two parts. One deals with the long term; the non-current assets and liabilities, and these are managed with reference to the airline's financial strategy and the decisions on the buying or selling of long-term assets. The second part is the management of current assets and liabilities. These two parts come together when the airline produces its short-term and long-term plans and forecasts.

The financial strategy, adjusted for agreed short-term variations, is used as a guide for determining non-current assets and liabilities. The measures used for current assets and liabilities, also adjusted for short-term agreed variations, are used to produce figures for these headings. When these figures are put together, they produce the airline's planned and forecast financial position. Forecasting is discussed in Chapter 6.

Management of an airline's balance sheet is not an easy exercise, but it is an important process and in times when trading is difficult, it becomes vital. Ensuring that the amount of money in an airline's balance sheet is used efficiently is important to every department in the airline. It is not an issue for the Finance Department alone. Table 5.2 shows that the amounts of money involved in operating an airline are significant, and that they need to be managed to ensure the airline is efficient and profitable and has sufficient funds to continue its operations.

6 Managing the airline's money

The efficient management of its cash and borrowings is a vital function in an airline. It is every bit as important as having a high level of customer satisfaction. An airline's cash-flow involves a significant amount of money even if the airline is considered to be 'small'. An example to put these comments into prospective: if an airline has one leased aircraft flying four flights a day with 70 passengers on every flight, each paying a fare of USD 300, the total cash received each day will be USD 84,000 or USD 30,660,000 a year; assuming the airline makes a profit of 5% on revenue, its operating costs will be USD 29,127,000 in a year. This 'small' airline has a total cash-flow, in and out, of just under USD 60,000,000 a year, or USD 164,000 per day; in addition, the profit needs to be either re-invested or distributed to shareholders.

Included in cash management is the management of the airline's 'financial resources', which is its ability to borrow funds at a reasonable cost whenever it needs them. The airline does this by achieving the objectives of its financial strategy and not exceeding the limits the financial strategy imposes.

Virtually everyone in an airline is involved to some degree in managing an airline's current and future cash-flow. Staff need to be aware, not only of what is happening currently in the airline, but also of the possible effects on the current and future cash-flow of each decision they make. Most decisions have an effect on current and future cash-flows. As an example, if a decision is made to increase the amount of emergency stocks held at a particular airport in order to reduce the chances of a technical delay, this involves using part of the cash inflow from passengers to buy the extra stock. The only circumstances where cash-flow is not needed is if there is surplus stock elsewhere in the airline's network and, if this is the case, it shows that excessive stocks have been held, hence from a cash management point of view, too much money has been 'invested' in emergency stocks.

The Treasury function within the Finance Department is usually solely responsible for managing the airline's cash and near-cash resources, in its home country and internationally, but all decision-makers in the airline have a significant influence as well.

In the airline's statutory reports, the total of cash and similar funds is often described as liquid funds, and for this reason cash-flow management is sometimes called 'liquidity management'. Added to the normal difficulties managing the daily cash movements, inflows and outflows, the Treasury team must be able to react to occasions when substantial funds may be needed urgently, e.g., if a flight is diverted to an airport the airline does not usually fly to, funds will be needed for landing and parking charges, passenger expenses and possibly engineering costs.

Statutory reporting of cash-flow

Cash-flow is sufficiently important to every company that it is reported separately in statutory reports. There are usually two parts to the reporting:

- A note which shows the detail of how the operating profit becomes the operating cash-flow generated. The statement is often referred to as the 'reconciliation of operating profit to cash generated from operations.' A reconciliation is a process which compares records prepared from the same or similar sources to prove they are correct by listing the differences. A simple example of this is when an individual compares the entries on their bank statement with the information on their own record of transactions (cheques drawn and cheques paid-in).
- An example of the statement of operating profit and cash-flow for a fictitious airline is shown in Table 6.1.
- A statement of the actual total cash-flows for the period. An example of the statement of cash-flows for a fictitious airline is provided in Table 6.2.

Table 6.1 Reconciliation of operating profits to operating cash-flow

Reconciliation of operating profits to operating cash-flow	
	USD m
Operating profit	114 *(figure shown in P&L account)*
Non-cash charges	*(items not involving cash payment)*
Depreciation	14
Amortisation	1
Provision for doubtful debts	1
Operating cash-flow	130 *(change in operating cash-flow for the period)*

Profits and cash or money are not the same thing, although in casual conversation the terms are often treated as being interchangeable. For example, the question 'How much money did XYZ Airlines make last year?' in most cases really means 'What was XYZ Airlines' profit last year?' 'Profit' is the gain arising from the operation of a business. In the case of an airline carrying passengers and cargo, the amount is the revenue earned where the service has been completed, less the costs incurred for those completed services. 'Money' or

Table 6.2 Statement of cash-flows

Statement of cash-flows	USD million
Operating	
Operating cash-flow	130
Interest received	1
Interest paid	(3)
Tax paid	(2)
Net cash inflow from operations	126
Investing	
Proceeds from the sales of assets	65
Purchase of assets	(85)
Dividends from associates	2
Cash outflow from investing	(68)
Financing	
New loans	65
Loan repayments	(12)
Dividends to shareholders	(13)
Cash inflow from financing	40
Increase in liquid funds	98
Liquid funds at the start of the year	35
Liquid funds at the end of the year	133

'cash' is the means of exchange we use every day; either physical money such as coins and notes, or electronic money such as in bank accounts.

The reconciliation of operating profits to operating cash-flow removes from the operating profit before interest and tax, all non-cash costs to produce the cash-flow from operations. 'Non-cash' costs are costs which do not involve any movement of cash and are estimates (e.g., an estimate of the value of debts due to the airline that may not be paid) or calculations (e.g., depreciation, which is an estimate of the reduction in the value of assets used in the business during the period). The reconciliation starts with the 'operating profit' and finishes with the 'cash-flow from operations'.

An airline has other cash-flows, such as buying and selling assets, repaying loans etc., and these are included in a 'statement of cash-flows', which is part of the airline's statutory reports and summarises the cash movements for the airline's whole operation. This statement starts with the figure for the 'cash-flow from operations' from the reconciliation of operating profits to operating cash-flow. It then shows the other cash-flows, in and out, classified by their nature,

- 'Operations' (from operating the business).
- 'Investing' (buying/selling assets, including investments in other companies).
- 'Financing' (loans received, loans repaid and dividends to shareholders).

The final figures at the end of the statement of cash-flows show the total change in the airline's liquid funds, cash, deposits, etc., during the year and also the total at the start and end of the year; the figure for the end of the year agrees with the figure for liquid funds in the balance sheet.

With these two pieces of information, a reader can understand what has happened to the airline's cash-flow during the period. In many respects the cash figures in the statements are the easiest to understand because they deal with actual cash movements, both in and out, and are similar to the cash transactions each person deals with every day, although the figures are much larger.

Management's reporting for cash-flow and its management

The cash-flow information included in the statutory reports is simply a historical summary. It is important information which can reveal trends and can indicate possible future problems, but the reports do not include any information on future cash-flows. The company's statutory reports are required to include a discussion of the company's future prospects, but any discussion on the company's financial future usually focusses on profits and its financial KPIs.

The primary objectives of cash-flow management are to,

- ensure that funds are available to the airline when they are needed
- ensure surplus funds earn the maximum amount consistent with the airline's risk tolerance
- minimise the cost of borrowings
- work to achieve the airline's financial strategy and make transactions which are within the airlines approved limits
- report the airline's cash and financial position regularly and promptly.

Airlines are companies which naturally have to look into the future, both near-term and long-term. Most of the key decisions take a long time to implement and it may be years before the full benefits are realised. There are decisions which can be made now but cannot be implemented until some time in the future. Examples are:

- The desire to fly to a new destination, which may require the negotiation of traffic rights and airport landing slots.
- Delivery of an aircraft which depends on the availability of space in the manufacturer's building programme or in the lessor's fleet.

These types of decisions require significant amounts of money to be available and the timing is very important. Virtually every part of an airline's medium- and long-term plans will require additional money to be invested and frequently the appetite for money exceeds the amount available.

An airline's cash-flow is managed day by day, and with a long-term view against the background of the airline's latest operating plan and forecast results.

There are at least three documents which are particularly important to the Finance Department. These are the airline's:

- Short-term plan. This may be called a 'budget' or 'plan' and usually cover a period of 12 months. The plan will include the amount and timing of expected operating income and expenditure, balance sheet transactions (e.g., purchases of assets and their funding) and payments of dividends. This information is used as part of the Treasury's short-term cash-flow management.
- Short-term forecast, which is similar to the information in the budget, but the information is revised to bring it up to date with the latest decisions and timings.
- Long-term plan, perhaps covering five years, showing future operating income and expenditure and balance sheet transactions analysed into years or some other convenient period.

The detailed information in each of these reports will be integrated into the Treasury's schedule of long-term commitments. This schedule shows all payment commitments the airline has for its major borrowings (e.g., leases, loans and instalments for aircraft purchases). The schedule will cover all the airline's commitments that have been made, and can easily cover 15 years or more. The reason the Treasury keeps its eyes on the potential future financial results is that arranging major financing transactions, such as issuing more shares or issuing bonds, needs to be planned and can take many months to put in place. The availability and the cost of finance changes as economic circumstances change. If the Treasury foresees the need for a major financing in (say) three years, it may recommend arranging the finance well in advance if their judgement is that the financing market is favourable for borrowers.

Producing the financial projection integrating the latest prediction of financial results with the schedule of existing commitments shows whether there may be potential cash-flow problems in the future. It may be that the payments for assets and loan repayments required in one or two years' time are exceptionally high, and action may be needed to smooth the fund outflow by perhaps taking a short-term loan. In most cases, the long-term financial projection will result in a list of items to keep under review. The short-term plan and the short-term forecast will also show whether there is likely to be a change in the airline's currency inflows and outflows; changes could arise from flying new international routes. Significant changes may mean that the Treasury will need to change its borrowing tactics or consider arranging forex sale contracts. These reports will also show whether the airline is likely to remain solvent during the foreseeable future. Without all this information, the Finance Department will find it difficult to manage the airline's funds and financial position. Making arrangements to ensure the company's borrowing, currency inflows and outflows and overall cash-flow is managed in the most efficient way can often take several months to organise.

The outcomes from the short-term and long-term plans and the short-term forecast which particularly interest the Finance Department include:

- An indication as to whether there are likely to be any short or long-term funding problems. If there are, it will lead to a discussion on how best to resolve them.
- Whether there is any danger of the airline being insolvent at any time, even for a short time.
- Whether the airline's financial position will be in accordance with its financial strategy.
- What the airline's forex position will be, which currencies will be in surplus and which in deficit.

The difficulty and complexity of the Treasury's work depends on:

- The nature of the airline's business. If, for example, an airline's business is highly seasonal, it is likely that maintenance work will be deferred to the low season. Therefore, the airline must accumulate funds during the busy season to provide funds for the time when revenues are lower but costs are higher.
- The complexity of the business. If the airline has many overseas offices and/or significant investments in subsidiary companies and perhaps provides services to other airlines, gathering together all the information needed to produce one all-embracing fund position may be difficult, and it may be more efficient to keep the cash-flows separate and ensure each is efficiently managed.
- The scale of the business. Clearly it is easier to monitor the fund position of an airline operating a small fleet of (say) five or six aircraft, compared with an airline with 50 or 60 aircraft, but the same degree of care is needed. Although the smaller airline's figures are smaller and easier to understand, the scale of the business is sufficiently small that a major deviation from the forecast may produce a devastating change to the airline's cash-flow, whereas with a larger airline a large deviation in one area may be compensated for by a favourable deviation in another area.

Plans and forecasts

In addition to these financial outcomes, the plans and forecasts also indicate to the Finance Department whether the information and format of the financial management reports to all the operating levels in the airline is still appropriate. Changes are not needed very often, but there will be times at which a revision will help the airline's management understand what is happening. For example, a small airline may get regular reports on the profit or loss and cash-flow produced by each flight, but as the airline grows, this amount of information may be too much for the more senior management and the reporting may be changed to the profit or loss and cash-flow from each route, with flight profitability monitored at a lower level of management less frequently.

The Finance Department is not a passive observer in the process of producing the plans and forecasts. It provides some of the essential information used in the preparation of plans and forecasts, primarily:

- The level of interest rates for the various loans the airline has or is planning to have, as well as the amount of interest income likely to be earned on surplus funds.
- Forecast forex rates for all foreign currency transactions, whether for income or expenses.
- Forecast inflation rates for all the countries the airline operates to or has offices in, split if necessary into categories (i.e., staff costs and property costs).
- Tax rates and the basis of calculation for each country where the airline pays tax.
- The historic relationship between balance sheet items like the amount invested in engineering stocks and the airline's engineering activities. These will be the measures or ratios used to monitor the current assets section in the balance sheet, as discussed in Chapter 5.

Routine management

Day-to-day management of an airline's cash-flow relies a great deal on information flowing into the Treasury from operating departments and from any operating offices the airline has in other countries. The Treasury's primary objectives include ensuring that there are sufficient funds available to meet payments when they are due and that any surplus funds, whether short-term or long-term, are invested to earn interest. Achieving these twin objectives means that the Treasury will seek to impose some routine on the payment of suppliers' invoices and capital sums, and also seek to ensure that all debtors pay on time.

It may be that the Treasury will seek to make payments to suppliers regularly, once or twice a month, so the Treasury's workload can be better organised, cash-flow is easier to control, and creditors will have confidence that their invoices will be settled 'on schedule'. If suppliers' invoices are usually paid once a month in one batch this will be the easiest option for the Treasury, but if all the invoices approved for payment are paid on the next payment date, it could mean that some invoices are paid more quickly than the credit period granted by the supplier, and this may be seen as a less than efficient management of cash. Some system has to be established and followed. Usually staff salaries will be paid regularly, perhaps once or twice a month. Staff salaries are a significant cost, particularly for flying staff. The amounts due to these staff can also be difficult to calculate because of the complexity of industrial agreements, and it is important that figures are finalised promptly to enable the staff to be paid on time. If urgent payments are needed, short-term deposits may be called back, or the airline's overdraft facility used. The Treasury will take whichever course produces the greatest benefit for the airline.

Controlling the repayments of borrowings is both easier and yet more difficult. They are easier because the dates for the payment of interest and repayment of the loan dates are agreed when the loan commitment is made. Hence the information is known not only well in advance, but also for the whole period of the commitment, and this can be as long as 10 to 15 years. The main difficulty is to arrange for the individual repayments of interest and loan repayments to fall on the same day of the month, or to make it so that they are at least grouped closely together. Managing the dates of payments is done during the negotiation of the facility, whether it be a lease agreement, loan agreement, instalment payment for an asset being purchased or for a service to be provided. The staff in every department when negotiating the contracts will be encouraged to discuss and agree the supplier's payments terms with the Treasury, so that payment dates are agreed which follow the airline's payment system.

An airline's fund inflows split into two parts: systematic receipts and irregular receipts. The systematic receipts probably cover most of the airline's receipts and this applies equally to Low-Cost Carriers and Full-Service Carriers. A Full-Service Carrier's revenue will be partly collected through a central collection system in each country, which collects ticket or cargo revenue from customers and pays the airline the amount collected (less a fee), and partly from individuals and companies paying through credit card companies. In addition, there will be amounts due from and to other airlines for multi-airlines' journeys, which are settled regularly through a separate clearing system. Low-Cost Carriers tend to collect their revenue mainly through credit card companies. The payment dates for the amounts due is agreed when the airline signs the agreement covering the collection. The Treasury will monitor the prompt receipt of the income and, if there is a delay, refer to the department which deals with the contract.

In addition to these regular revenues there can be many irregular receipts which become due, and these can be generated from any area of the airline; examples are,

- charges to third parties for advertising in inflight magazines
- fees for performing the ground handling for other airlines
- re-charging the cost of spares made available to other airlines under a spares' swapping agreement
- excess baggage charges collected at airports
- inflight sales
- ticket sale at airports and the airline's offices.

The list seems to be endless.

The total amount charged by even a small airline can be millions of USD. It is important to know the absolute amount that is due so that the collection can be incorporated into cash forecasts, but the essential day-by-day action is to collect the amounts due on time and, when this does not happen, to highlight the fact and take action to collect any amounts that become overdue. The responsibility for collecting these irregular collections rests with the department which made the arrangement and issued the invoice, but the Treasury needs to be kept in the picture because they manage the funds once received. It is inefficient for the Treasury to uplift a deposit or to increase its overdraft at a bank in order to make payments, only to find that a substantial invoice from a debtor is settled the same day. The result will be that the chance to earn interest or avoid an overdraft interest payment will have been lost.

The day-by-day and month-by-month control of cash-flow is essential to an airline because of the volumes of money handled. It is particularly important to airlines which are developing and expanding. It is too easy for management to concentrate on developing the business and gaining passengers, but to forget about the delays in receiving the money. Even where passengers pay their fares using a credit card, there is a delay between the transaction and credit card company paying the airline. The payment date will have been agreed and included in the service agreement, but that date will include an allowance for the credit card company to process the transaction. Good cash-flow control is not a 'nice to have' facility to be introduced once an airline has become established, but an essential system to have in place from the very first day an airline, of whatever type or size, is established.

Policy matters

There needs to be clarity and a well-understood arrangement between the airline, its customers and suppliers, for the handling of amounts owed to and by the airline. Informal or ad hoc arrangements make cash-flow management more difficult, and can lead to confusion and misunderstandings between the company and its customers or suppliers. People tend to be sensitive about money that is due to them, and a well-regulated system can avoid disputes and make for smooth processing. To achieve these objectives, policies for dealing with collections from debtors and payments to creditors should be established and adhered to. Clear Debtors and Creditors Policies, which are workable and understood by all staff, will help to minimise the amount involved in working capital and assist in the management and control of working capital as discussed in Chapter 5.

The airline's policy for dealing with the collection of amounts due from debtors and the payment of amounts due to creditors will be set by the airline's Finance Committee because it is responsible for establishing a procedure for monitoring financial relationships. Copies of the airline's 'Debtors Policy' and 'Creditors Policy' will be distributed to every operating department to guide them in their negotiations with customers and suppliers. Compliance with both of these policies will be examined periodically by the airline's Internal Audit Department, and a report will be given to the departmental managers involved, the airline's Board of Directors and the airline's statutory auditors.

The detail of each of the policies will have to be varied to comply with the local laws and/or practices in each of the countries in which the airline operates. Also, each airline's management will have its own idea as to what is acceptable business practice for the airline. These two factors, local laws and practice and the airline's standards, need to be borne in mind when looking at the basic headings in each of the policies.

Debtors' policy

The main items in a debtor's policy typically are as follows:

- Mandating which departments are responsible for deciding the amount to be charged to customers for services and goods, and also for collecting all amounts due. Usually this is the department which negotiates the deal or service agreement.
- Stating the minimum information about each debtor, which must be obtained before credit can be granted. This section will also state how frequently the information should be brought up to date and who should review the information.
- Setting a basis for deciding the maximum amount that each debtor may owe the airline together with a procedure for varying the limit, either temporarily or permanently.
- Stating what credit checks on debtors and potential debtors are required and how frequently they should be updated. The checks may include a bank reference; it is likely there will be different levels of checks depending on the probable total amount of the debt due to the airline.
- Setting a maximum time limit for credit given to debtors. This is likely to be a standard time, but there may be exceptions for certain debts (e.g., refunds of any over-paid taxes or insurance claims).
- Establishing a procedure, action and timing, for dealing with debts which are not settled within the agreed time limit. These may range from sending standard letters asking for payment to legal action
- Establishing whether interest should be charged on amounts which are 'overdue' (i.e., unpaid after the maximum time limit). In practice, it is often difficult to enforce payment of interest on overdue amounts where there is a continuing commercial relationship.

Even for established airlines, managing debtors requires vigilance. Table 6.3 shows the overdue debts at the year-end as shown in the statutory reports of two Full-Service Airlines. The figures show that despite having established systems and years of experience, some debts remain overdue and although the figures are not significant when compared to the airline's revenue, the total sums of money are significant. Just think what assets could be purchased for about USD 60 million.

Table 6.3 Overdue trade debts

Overdue trade debts

	Airline 1		Airline 2	
Year-end	30 June 2017		31 December 2017	
Currency	Local currency million	USD million equivalent	Local currency million	USD million equivalent
Past due 1–30 days	60	44.6		
31–120 days	13	7.4		
Over 121 days	13	7.4		
One to three months			395	50.6
More than three months			126	16.2
TOTAL		59.4		66.8

Note that different periods are used for showing the age of the debt.

Creditors' policy

The policy for creditors is not too different to that of the debtors, but views the situation from the 'other side':

- Mandating the management level, which has the authority to agree contracts with third parties and, if appropriate, setting monetary limits for each manager or department.
- Stating the information, which can be given to a prospective supplier of goods or services as part of their due diligence.
- Stating a preferred credit period to be granted by the creditor and advising the procedure for amending it.
- Setting a maximum amount which can be owed to a creditor. This includes banks and financial institutions.
- Establishing the regular payment dates for payments to creditors.
- Setting a limit for the time to resolve any issues on amounts invoiced and, if necessary, stating a procedure to settle significant disputes. This is an important part of the policy because it influences the airline's reputation in the industry and community.

Exchange rate policy

The airline's approach to managing the risk of losses from changes in the foreign currency exchange rates is part of managing an airlines money. This is certainly the case for an airline with an international network, i.e., one that has earnings, costs, creditors and debtors in different currencies, but even domestic airlines can have exposure to the USD for aircraft spares and fuel. Changes in the rate of exchange between foreign currencies and the airline's home currency can have a significant effect on the airline's profits and cash-flow. For example, if an airline has debts in a foreign currency, and that foreign currency strengthens against the airline's home currency, the airline will suffer a loss. Chapter 4 discussed the techniques that can be used to control the effects of changes on the airline, but prior to applying the techniques, the airline's attitude to forex risk has to be established. In the absence of a policy, the Treasury section will not be able to manage forex risk.

Changes in foreign currency exchange rates can influence more than just the amount of cash flowing into and out of the airline's bank accounts. In addition, they can affect the airline's essential business:

- The business; a significant change in the rate of exchange between the airline's home currency and one used in a foreign market may cause the number of passengers travelling and/or the amount of cargo shipped to change significantly.
- Business values; if the airline has operations in foreign countries (e.g., flight kitchens), the value of those operations in the airline's balance sheet will change with movements in the forex rate. It is possible for the Treasury to hedge any movement of funds to or from the foreign operation (e.g., profit remittances or additional capital) to limit any adverse effect. There may also be other foreign currency assets and liabilities in the airline's balance sheet which can change (e.g., overseas debtors or rent deposit for overseas offices).

An airline should have a view and stated policy on foreign currencies. In Chapter 4 there was a discussion on the financial techniques that can be used to manage forex risks once they have been identified, but before the techniques can be used the airline's management should issue clear guidelines on how to deal with foreign currency risk. The starting point is to decide which forex risks need to be managed because an adverse change will have a significant effect on the airline, and which forex risks can be ignored at present. The criterion used to decide is usually the effect on the airline's profit. In this case, a decision is needed on the maximum amount of adverse change to profits which can be tolerated. If a volatile foreign currency is a very small proportion of an airline's income or costs, it is probably not worth trying to manage the loss from fluctuations, but management must decide on the cut-off point and must also ensure there is a system for monitoring the position to ensure the decision remains valid. Quite often the cut-off point is set at something like a maximum effect on profits of 0.1% or more, but management must form its own view. The evaluation process is called 'Value at Risk', often known by the initials 'VaR'.

The airline's statutory report will include an assessment of the major currency risks in the section which discusses financial risk management. An example of a currency risk assessment is presented in Table 6.4.

Table 6.4 Sensitivity analysis for foreign currency risk

	Increase/(decrease) in the result *USD million*	*Increase/(decrease) in equity* *USD million*
Euro	1,076	(448)
Pound sterling	(130)	190
Australian dollar	25	175
Net change	971	(83)

A 5% appreciation of the airline's home currency against the airline's major trading currencies would have resulted in the following changes in the airline's result for the year and a change in the value of the airline's equity.

When reading this information, it should be remembered that the evaluation is usually only of the effect of changes in forex rates for the major currencies in the airline's current cash-flow and foreign currency assets and liabilities. Generally, there is not a discussion of the possible wider economic effects of any change in a forex rate (e.g., whether a change is likely to result in an increase or decrease in passenger numbers or cargo carried, or whether a change will produce the need to change suppliers or the way the airline is organised).

The foreign currency policy options for the airline are the same as for any other business dealing substantially in foreign currencies:

- Do nothing; the assumption underlying this is that forex risk is just one part of the whole package of risks inherent in the airline's business and the company will simply accept the effects of any rate changes. This is a perfectly valid approach, provided it is a conscious decision having considered the possible effects on the airline and the costs of hedging, both for the transaction and in management time. Simply failing to consider the possible effects of exchange rate changes is not really a sensible option.
- Hedge everything; this uses the approach that the airline's profit arises from completing a transaction (e.g., carrying passengers or cargo) which involves certain risk, and that all other risks must be hedged. This approach must take into account the costs of hedging and accept that it eliminates the possibility of 'windfall profits' (i.e., unexpected gains).
- Hedge just part of the risk; this is the approach taken by many airlines and is a combination of the other two. It involves doing hedging transactions in those foreign currencies which could have significant adverse effects on the airline and ignoring the rest. The policy may also contain guidance on what action should be taken when a foreign currency rate is changing or is forecast to change more rapidly than normal market fluctuations. With this policy there must also be a system to monitor changes in the significance of each foreign currency to the airline.

Whichever policy is adopted will depend on the airline's management attitude towards risk and this does tend to vary company by company.

Successfully managing an airline's money involves specialist techniques, and to function efficiently the Treasury needs up-to-date information from operating departments. The management of cash-flows is a vital part of an airline's operations. Although the Treasury's operations dealing with financial risks may seem to be an extra service, they are in fact designed to protect the airline from unexpected financial shocks.

7 Planning and forecasting

Considering the future and the airline's place in that future is an essential part of running an airline. Such important topics as negotiating traffic rights, adding to the aircraft fleet, introducing a new passenger product (e.g., new seats and service), developing a computer system to make check-in easier or re-organising an airline's finances all take a lot of time and require an airline's management to spend a significant time considering the airline's future development.

It is worth noting that in many jurisdictions company law or stock exchange regulations may require that a company's statutory reports should include the Board of Director's current view of the company's future in its review of the business. The requirement is not a forecast of profits, but a discussion of the main trends and factors that could affect the company's future and what they plan to do. There is not a requirement for a profit forecast or for the directors to reveal commercially sensitive information. The existence of commercially sensitive information that is private to a company is acknowledged. Information such as trade secrets and deals being considered or negotiated need not be disclosed if revealing them would damage the company, and the judgement on what may damage the company rests with the Board of Directors. Some companies include a page tabulating the company's opportunities, challenges and priorities. The discussion of the future often centres around the company's Key Performance Indicators often referred to as KPIs or Success Measures and possible changes that are of interest to investors and other stakeholders.

Key Performance Indicators

KPIs are factors which measure the vital parts of the business. They are, for example, the pieces of key information the chief executive of a business needs regularly to understand what is going on in a business. There should not be too many KPIs, but estimates vary on what the ideal number should be. Some believe that it should be possible to report all KPIs on one side of an A4 sheet of paper. The word 'Key' is important, as only essential and important matters should be included in the set of KPIs. Usually there will be KPIs which measure a company's finances, operations and development and these will be expressed as,

- purely financial measures, (e.g., return on equity)
- purely operating measures, (e.g., percentage of on-time departures)
- Combined financial and operating measures, (e.g., average revenue per passenger).

To be effective, KPIs need to be,

- relevant, i.e., measure an essential part of the business or a function which temporarily needs management's attention
- focussed on important, almost existential matters
- without bias and able to show both bad and good results objectively
- used consistently for each period; if it is necessary to change the way a KPI is calculated, the figures for past years should be re-calculated using the new basis so that comparisons remain valid and any trend can be seen
- accepted by the airline's management as effective and efficient measures.

It should also be agreed when a trend in a KPI is an acceptable trend. This can be done by setting a target. For example, if a KPI is 'training hours per staff', it is not clear just from the KPI whether an increase is good or bad. Does an increase mean that staff are becoming very highly trained and efficient or too highly trained (i.e., acquiring knowledge which is not useful)? Alternatively, does it indicate that staff being recruited do not have the required level of training and experience and the airline must fill the gap?

There are some who argue that the remuneration of senior managers should be related to achieving pre-agreed levels of KPIs.

In many respects, planning is more important for a small start-up airline. Adding one aircraft to a fleet of four aircraft probably takes more detailed planning than adding an extra aircraft to a fleet of 100, as it is probably more of a risk to the smaller business.

Financial planning is part of the airline's overall planning and strategy process. An airline's ability to raise further funds when it needs them, at an acceptable total cost, should be considered as a vital resource for the organisation, in the same way as aircraft and the expertise of staff are resources. The airline's success and the progress towards achieving goals will be measured partly in money. These factors mean that the Finance Department has an important role and responsibility in planning, forecasting and strategy.

Chapter 6 mentions three documents which are important to an airline's treasury function, being:

- short-term plan
- short-term forecast
- long-term plan.

Essentially, the short-term forecast is an update of the original short-term plan. These are three documents that look forward, covering different periods, and are important not just to the Treasury section but the whole organisation. The forecasts guide the Treasury in how long to invest surplus funds, and whether there is the potential for a financing problem in the future, as well as helping them plan when to arrange finances for the airline. Short- and long-term projections help other departments in a similar way, highlighting potential problems and indicating possible opportunities.

Running an airline successfully requires management to take a long-term view of how the airline will develop, setting long-term objectives, and measuring actual progress against these objectives. It is easy to discuss the outline of the process of looking into the future, but it is difficult to describe the detailed steps because each airline takes a different approach, uses different techniques, and has different names for the steps in the process. There are many permutations. For example, one approach is to produce one document which starts with a short-term forecast in detail, quarter by quarter for (say) a year, then changes into a long-term plan with summary figures for each year. Some airlines, in addition to short-term and long-term plans, may also produce a medium-term plan as a separate document. In all the approaches, the plans need to be linked together in some way to show a consistent view of the likely results. The amount of detail in each is likely to differ, as the plans covering the longest period are likely to be shown in less detail.

When discussing the always hazy long-term future it may be easier to use words rather than attempt to project what the actual figures will be. For example, it is often easier to note and understand a description of the future than to try and calculate the revenue figures for five years in the future. Statements like 'the airline currently has a 15% market share of passengers leaving its home airport'; 'the XYZ tourist authority expects the number of departing passengers to increase steadily, possibly by between 7% and 10% each year, in the next 5 years'; and 'we plan to keep our current market share' are easy to understand and act upon. These sorts of statements in the long-term plan give those operating departments who need them a basis to calculate a range of possible figures to create an idea of how future demand may be met. They also eliminate the need to make small adjustments to the forecast figures if there are periods when there is a small change in the underlying assumption (e.g., passenger fares are expected to increase by 1.1% rather than 1%).

The basic problem with planning and forecasting is that the processes deal with the future, something which is unknowable. This is an obvious comment, but sometimes businesses need to be reminded that the future is unknown and may actually be quite different to their current view of it. Although it may be possible to make inspired guesses of what might happen within an airline, there are many external factors beyond management's control or even beyond their knowledge. An example is some new technology like automated check-in, which significantly changes how passengers are treated and the associated costs. These factors can influence the actual outcome, regardless of the quality of management's internal thinking and planning. It is not possible to avoid uncertainties when constructing a plan or making a forecast. The amount of intellectual effort and detail put into the preparation of a plan or forecast must be balanced with the likely accuracy and usefulness of the final plan or forecast. For example, there is little point spending six months producing a detailed plan for the following year if there is a high likelihood it will be out of date when it is issued. A plan is not an exact, well-defined picture of the future but an informed guess based on a whole range of assumptions. Some of the information will be known with a high degree of certainty (for example, the future composition and capacity of the airline's fleet over five years), but even this may change if external factors dictate that a review is needed of the capacity (i.e., number of seats) in order to remain competitive. This lack of accuracy can make it difficult for the airline's management to develop options for meeting targets.

To be most useful the airline's long-term plans should concentrate on those factors and elements which have a vital effect on the airline's business. All of these vital factors should be included in the airline's KPIs. These are those measures which most effectively indicate the airline's progress towards achieving its strategy and are those measures which are key, that is, 'essential' to running the business successfully. Planning and forecasting is an area where too much data can cause confusion, and a very detailed output may lead the reader to conclude that the projections contain more certainty than they actually do.

The external factors which can influence an airline's financial and operating results can change quickly. For example, a competitor may announce a new, upgraded product or establish a new route which increases competition. A change in the economy in a major market may produce an increase or decrease in passengers and cargo.

Approaches to planning and forecasting

There are many ways to produce a plan or forecast. The range of approaches is wide:

- Projecting past trends for revenue and cost items into the next period, together with historic ratios for balance sheet items.
- Projecting revenue and costs based on best guesses for the following period and maintaining balance sheet ratios.
- Reviewing all sources of revenue and cost items and the structure of the balance sheet.
- Setting future targets based on KPIs and translating them into monetary amounts.
- Making the development of a company plan into a continuous process. This can be achieved by reviewing the detail of each of the airline's key areas: income, costs and balance sheet items, in sequence throughout the year, and developing a plan for them for the following 15 months; hence,

for example, flight operations in each January, sales and marketing in each March, engineering in each May, etc. This approach means the whole airline is reviewed once a year. After each of the reviews a plan for the following (say) 15 months is produced. This approach spreads the workload evenly over the year, and ensures the whole airline is reviewed regularly each year and the whole airline has a plan.

There are variations of these approaches. The whole subject of short- and long-term planning has been under review and discussed for many years, and one single approach has not been universally adopted in aviation. The only consistent conclusion is that short- and long-term plans and forecasts are an important part of running an airline successfully.

Some airlines have developed a model of the factors influencing their results and operations, and use this to produce plans, forecasts and evaluate significant changes. The model links together all of the effects of a significant change in its operation. If, for example, there is a decision to expand into a new market in response to competition, there will be effects on the utilisation of the airline's fleet of aircraft and this may produce the need for additional aircraft, which, in turn, may mean more borrowings that will have an effect on the airline's debt/equity ratio. In addition, more cockpit crew and ground staff will probably be needed and, in addition to the training costs involved, the increase in staff numbers may have an effect on the airline's pension scheme. The model links all these factors. Producing such a model can be quite difficult and is the sort of project usually undertaken by a large and mature airline. An advantage of having a model of the airline's business is that the management can also use it to quickly evaluate the effect of a significant change, so that a response can be agreed.

Another approach to producing a plan is for the airline to compare the position shown by the figures of its current KPIs with the figures for the long-term target's KPIs, and decide what improvements need to be made and what should be achieved in the period of the plan.

Whatever approach is used, the cost and effort of producing each plan and forecast must be balanced against the accuracy, hence the value, of the output. Those airlines using a complete computer model will probably be able to produce a number of results: 'best', 'most likely' and 'worst', and if that is done, a decision must be made on which set of figures will be used as that standard against which to measure actual performance. Others airlines may produce just one report, but with a note of the effect of significant changes to the most influential factors, such as:

- interest rates
- forex rates
- passenger revenue
- cargo revenue
- fuel prices
- staff numbers and costs.

Table 6.4 in Chapter 6 shows the sort of note on sensitivity to changes in forex rates which is included in a statutory report. A similar note can be issued with forecasts and plans.

The KPIs used in the long-term planning, forecasting and evaluation of significant changes will include all of those expressed in the airline's strategy because the strategy is the current definition of what 'success' means for the airline. The financial KPIs are shown in the 'Measure' column of Table 2.1 in Chapter 2. In the financial strategy, the KPIs appear as maximum and minimum permitted values for each of the measures for each of the headings. The same ratios will be used in the plans and set as targets to be met on the road to achieving the airline's desired position. If forecasts show that progress in the ratios is in line with the plan, then it is likely that no action will be taken, but if there is a worsening trend or a worrying trend has not changed, remedial action will be needed.

Using Key Performance Indicators

Using KPIs should make it easier for everyone to understand what is happening in the business and what the airline wants to achieve. Not everyone will want to master the information in the airline's statutory reports, but they should be able to understand the KPIs because all of them relate to the business and one or two will directly relate to the part of the business they work in. In order for KPIs to succeed as a means of communication they need to,

- Be consistent; i.e., the same measures calculated in the same way every time they are reported. Some countries require that the essential KPIs are shown and discussed in the company's statutory reports and the figures and basis of calculation should be the same, regardless of whether the KPIs are being reported in the statutory reports or in internal management information.
- Cover every key area of the airline's operations and not just the financial results. Financial results are the outcome of operating decisions and these operations should be measured. Financial results are best understood when they are related directly to operations (e.g., revenue per passenger).
- Be understood by everyone in the airline. There are some KPIs which are used by most airlines (e.g., on-time performance, or average aircraft utilisation in hours per day), but this does not mean the airline cannot adopt its own, different KPIs if they are appropriate to its method of operating and/or its strategy (e.g., most airlines measure staff productivity with the KPI 'Available seat miles per staff member', but some also use 'Available seat miles per USD of staff cost').
- Measure all of the operations which are vital to the health and success of the business.

In addition to having KPIs for operation and the financial result, there should be KPIs relating to the environment and social effects of the airline's operations.

These will probably be required if the airline's shares are quoted on a stock exchange. A combined list may include:

- on-time performance
- market share by major market
- revenue per passenger, possibly by class travelled
- revenue per kilogram of cargo
- aircraft utilisation (hours per day)
- cost per passenger, or some other measure of capacity
- tonnes of greenhouse gas, 'GHG', emissions
- lost time injury rate (injuries per 100 staff).

The possible financial KPIs are in Table 2.1 in Chapter 2.

Comparison of the forecasts with the current plan and the longer-term strategy will highlight where there are likely to be shortfalls or problems in the future. They may also show those areas that are currently on the right course to achieve or even exceed objectives set.

Operating KPIs use both financial and operating information in order to make the figures more useful. Putting financial and operating information together makes the resulting ratio or KPI easier to understand when looking at efficiency. Trying to use only total figures, whether financial or other, to assess the position can be difficult and, in some cases, misleading. For example, if the actual total passenger revenue figures for two successive years are:

Year 1: USD 10,235,835

Year 2: USD 14,674,729

the year-on-year increase looks impressive at 43%. If the total number of passengers carried in the same two years was:

Year 1: 14,690

Year 2: 22,850

the increase would be 55%, which is also impressive.

In both years the total figures increased significantly, but it is not immediately clear whether the revenue performance in Year 2 was better than the performance in Year 1. If, however, the revenue per passenger is calculated, it shows that although more passengers were carried, the average fare paid by passengers decreased, and this may need to be examined to see whether it is acceptable. The calculation is shown in Table 7.1. The figures become even more useful if the actual performance is measured against the short-term plan as shown in Table 7.2.

This shows that the airline carried more passengers than planned, but at less revenue per head and that this result was worse than was expected. This example illustrates the value of putting total financial and operating figures into the context of actual operations and the airline's plan. The KPI 'revenue per passenger' was useful because it made the financial and operating figures come alive, and enabled staff to understand simply what had happened and whether the result was in line with what the airline's management planned for

Table 7.1 Calculation

Year	Actual		Revenue per passenger USD	Calculation
	Revenue USD	Passengers		
1	10,235,835	14, 690	696.79	USD 10,235,835 14,690
2	14,674 729	22, 850	642.22	USD 14,674,729 22,850
Change	+43.4%	+55.5%	-7,8%	

Table 7.2 Comparing with a plan

Year		Revenue USD	Passengers	Revenue per passenger USD
1	Actual	10,235,835	14,690	696.79
	Plan	10,300,000	14,820	695.00
	Variance	-64,165	-130	+1.79
2	Actual	14,674,729	22,850	642.22
	Plan	14,200,000	21,850	649.89
	Variance	+474,729	+1,000	-7.67

or desired. Seeing the reduction in the revenue per passenger will probably spur the airline's management to look closely at KPIs related to costs (e.g., possibly 'operating cost per passenger' or 'total cost per passenger').

KPIs are useful because they attempt to look at the airline's operation as a whole, rather than just its financial result, but there is a possible danger which needs to be avoided. It may be tempting to have too many KPIs, and too much data can confuse the reader and make it difficult to see important trends. There is not a recommended maximum or minimum number of KPIs a company should have, but the name itself gives an indication. The right number is the number which indicates what is happening in the key areas of the airline; no more, but certainly no less.

Comparing performance

The airline's KPIs and other ratios are very useful for making comparisons of actual performance with the airline's strategy, and for highlighting historic trends to give an impression of what has been achieved, but this is only part of the airline's picture. An airline is like any other business; it does not operate in isolation, but is in competition with other transport companies. In most cases the competition will be other airlines, but this is not always the case; the competition for some airlines is the train service, and this is particularly so when the airline flies on short routes between congested airports. Regardless of the type of company which is the competition, it can be helpful to know how well they are performing.

Most of the accurate information available on competitors can be collected from their statutory reports, although even here there will be some difficulties because of the different ways of presenting the figures reported and, because the statutory reports are financially oriented, all the operating information needed may not be reported. Nevertheless, making a comparison with the competition even using just the financial figures in their statutory reports can be very useful and constructive. It is also useful to look at the ratios and trends for other transport companies which are not yet in direct competition with the airline. For example, comparing Low-Cost Carriers with Full-Service Carriers with similar route structures may give indicators of what the future holds. Even a comparison of a competitor's total revenue and cost figures per passenger can be useful. Similarly, figures such as operating cost per aircraft, value of stocks per aircraft, or advertising as a percentage of revenue can all be helpful figures which may stimulate the airline's management to examine its own operations. The work of making comparisons with other transport companies falls to the department responsible for preparing the airline's plans. The Finance Department's assistance will be needed to examine and advise on the accounting policies and practices used by the other companies and it may be necessary for the Finance Department to re-work some of the figures in the other companies' reports to adjust them, approximately, to the accounting policies used by the airline. It should be remembered that the objective of making comparisons with others is to try and learn how their business is performing and to see whether there is anything the airline can learn from them. The purpose of examining and understanding the past it to help management make better decisions for the future. Trends can be as instructive as absolute figures or ratios.

There can be some value in comparing figures with companies in other industries which do not compete with the airline, but generally the total financial figures shown in the statutory accounts are not very useful, and it is necessary to try and find a way to get more detailed information. Examples of the way figures from another industry may help is by comparing the volume of water used by a bottling company to wash its bottles with the amount of water used by an airline washing its inflight cutlery and crockery, or comparing statistics on the response times of customer helplines with those of some banks. These are not financial figures found in a company's statutory reports, but there may be some reference in the report's text; if not, a direct enquiry to the company may be made. Any approach needs to be well organised and the comparison should be made with companies which are seen as having a high level of service. Organising comparisons with others is not always easy, but it can be very helpful. Comparing figures, particularly KPIs, with what others achieve can be stimulating, and may avoid the dangers of an airline only looking inwards at its own results and operations while ignoring what is happening with its current, and possibly future, competitors.

Tables 7.1 and 7.2 above show clearly that financial totals become more useful when they are combined with operational figures to produce an operating ratio. The tables demonstrate that it is easier to understand and more useful to compare the revenue per passenger rather than the total revenue or total number of passengers. The same logic applies to other parts of an airline's revenue (e.g., duty-free sales per passenger, per flight or per route). Similarly, it is easier to monitor costs when they are related to the operation that produces them and expressed as a ratio, as seen, for example, in Table 7.3.

Table 7.3 Examples of operating ratios

Year	Costs		Operating ratio	Operating figure
	Heading	USD	Per passenger	Passengers
1	Staff	2,341,000	USD 159.29	14,690
2		3,442,180	157.54	21,850
1	Operating	7,166,810	USD 487.87	14,690
2		12,295,429	562.72	21,850
			Per aircraft	Aircraft
1	Repair	1,710,000	USD 855,000	2
2		2,921,375	973,791	3

Recording and using these operating ratios can, over time, show very useful trends which can help the airline's management. It may be that the airline's management would rather see the ratios plotted on a chart rather than stated as figures in a report. For most people, it is easier to spot a trend from the figures in a chart than a tabulation of figures.

Table 7.4 Trends

Year	Per passenger USD		
	Revenue	Total cost	Margin
1	696.79	667.95	28.84
2	642.22	597.76	44.46
3	617.53	593.89	23.64
4	599.87	593.74	6.13
5	611.87	598.29	13.58
6	589.64	587.93	1.71

Table 7.4 shows a trend which is signalling to management that action needs to be taken to increase revenues, or reduce costs or both to improve margins. The figures show that management should already have initialled some action because the deterioration in the margin per passenger started in year 4.

Presenting the figures as ratios rather than as total figures helps to focus on trends and quickly highlights when something is not right.

Short-term planning

Strategy, long-term plans, forecasts, KPIs and comparisons are all useful tools in managing the direction the airline is taking, but there also needs to be some form of detailed short-term plan to ensure the airline's operation is still on course. This short-term plan is often called a 'budget' although there are many other names for the same document. Essentially it is a short-term plan which is related to the airline's current operating environment. The function of a budget or short-term plan is to:

- Act as a standard for monitoring and measuring actual financial and operating performance so that corrective action can be taken when it becomes necessary.
- Be a guide on whether the airline is progressing towards its long-term goals.
- Provide a way of assessing the effects of changes in operating circumstances and help when making decisions on the solutions to any adverse consequences of the changes.

A budget is a detailed plan for the operation of the whole company, specifying all the revenue and costs expected for the period, and organised department by department, function by function or responsibility by responsibility, together with statements of resources used (e.g., aircraft utilisation) and operating standards (e.g., on-time departures). Included in the schedules will be a list of all agreed expenditure on non-current assets, as well as any changes in current assets. The Finance Department uses this information to monitor and forecast the airline's financial position and produce a budgeted set of statutory reports. Usually the period covered by a budget is the airline's financial year.

In recent years, the short-term planning process in general has been reviewed by various business organisations and major businesses. There are many disadvantages to having a fixed budget for a long period like a year when revenue or cost factors may change significantly during the year due to many uncontrollable factors, such as forex rates or commodity prices. Nevertheless, some standard is needed to assess actual performance. The reaction to the problem by various businesses has produced a variety of ways of short-term planning, these include

- a 'fixed budget', which, once agreed by the Board of Directors, is not changed regardless of changes in circumstances
- a 'flexible budget', where the total figures are amended to reflect the actual number of passengers or cargo carried
- a 'rolling budget', which is revised every month or quarter to incorporate the actual figures achieved and adds an estimate for an additional month or quarter. Thus, there are always figures for the following year
- modifying a rolling budget to make more of a rolling forecast.

Regardless of its name or the exact process, the budget or short-term plan performs the same functions.

A particularly useful output is to compare the actual result achieved in a period with the budget for the same period and examine the reasons for any significant differences. This is often called 'variance analysis'. The 'variance' is simply the difference between the agreed budget and the actual result. Some of the variances can be automatically analysed into the major causes. For example, the variance for fuel costs incurred can be split into the amount due to the difference between the planned and actual price, and the difference in the amount of fuel consumed, but it is more difficult to automatically explain any difference due to changes in the airline's routes. Where there is a significant difference in any major revenue or cost items, the responsible manager needs to establish the reasons for the difference. The management's review process for comparing the airline's actual results with the budget varies greatly between airlines. The main reason for calculating the variances from the budget and explaining and reviewing them is to understand the figures, what is influencing them, and to see whether there are any lessons that need to be learnt so that performance can be improved.

Naturally, a very important account to analyse and review is the revenue. It is the most difficult to forecast and the most difficult to manage. Comparisons of competitors' revenue performance can be useful, if the information is available. Revenue is an important part of the budget and any forecast. In many respects, the level of revenue determines the maximum amount the airline can afford to spend on operating costs. Customers are the only source of income, and the money they pay is used to pay all operating costs. Yet revenue budgets and forecasts are very difficult to evaluate. The marketplace is a combination of many factors, and it is likely that the most influential factors are the business secrets of the airline's competitors. One of the strongest influences on the level of revenue an airline can achieve is the activities of the airline's competitors. At this point it is worth mentioning again the issue of knowing who an airline's competitors are. Not all airlines will be an airline's competitor, nor will all the competitors be airlines. For short-haul routes an airline's competitors may well include high-speed railways (e.g., on the route between London and Paris, and also Paris and Marseille). Not all Low-Cost Carriers compete directly with Full-Service Carriers, as they frequently fly from small, out-of-town airports and at differently scheduled times.

Usually revenue figures are examined by having detailed estimates for each route and each class of travel on the route, often using some form of ratio (e.g., revenue per passenger or per seat mile). These figures will be compared with whatever information the airline has about competitors' fares, as well as what has been achieved in the past. On a highly competitive route the airline may have to follow the market trend, but on a route with less competition, differences in fare levels may be tolerated by passengers because of other factors (e.g., age of aircraft, quality of service, or departure and arrival times). However rigorous the review of the revenue budget and forecast, it is still difficult to put it under as close an examination as the cost budget, the budgeted balance sheet and their forecasts. Revenue budgets and forecasts must include all ancillary revenue, such as from the sale of duty-free products, sale of food and drinks on-board, check-in charges or preferential seat charges. Every element needs to be identified so that the actual result can be compared to the estimate. In a highly competitive market, ancillary revenue can be very important.

Planning and forecast are vital roles in an airline and there is a wide range of techniques that can be used. The plans and forecast are used by the whole airline and the preparation should involve the whole airline.

8   Financial reporting

Objectives

A fundamental responsibility of the Finance Department is to report on the financial results and financial position of the airline. Preparing and communicating this information accurately, clearly and promptly is vital to the success of every company.

The main reason for the Finance Department recording the airline's financial transactions is to be able to show the airline's owners, staff and others what the financial operating result for a period has been and what the airline's current financial position is. The primary purpose of accounting is to be able to communicate the value and financial health of a company to the people and organisations that have an interest and commitment to the company. Therefore, it follows that the financial reporting which the Finance Department makes must be useful to, and understood by, the reader.

There is a general responsibility for a company to report on its operations and the responsibility for dealing with the financial reporting and the financial aspects of other reports falls to the Finance Department. It would be unreasonable to expect people and organisations to invest their money in a company and not be told how the money is being spent and what financial returns are being achieved. Similarly, staff invest their skills and time in the company and must be kept informed about what has been achieved financially and operationally; they also need to understand the company's current financial position, its assets, liabilities and cashflows.

Users of the reports

There are many people who pay attention to an airline's financial results and financial position, including:

- shareholders in the airline and potential investors
- financial institutions; current and potential lenders
- suppliers
- staff of the airline and their staff associations

- others, including government departments in the airline's home country and in other countries.

No doubt an airline's competitors are also interested in reviewing the information which is released, but there is no need to be concerned by this, as none of the reporting guidelines or statutory requirements obliges a company to reveal information which is commercially sensitive.

Figure 8.1 presents an overview of these external connections.

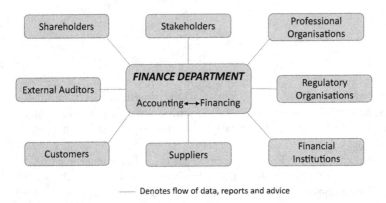

Figure 8.1 Connections 'outside' the airline

The groups are usually described collectively as 'stakeholders'. It is difficult to have a good understanding of exactly who is interested in understanding an airline's financial position or exactly what information they would like to have. The list of people or organisations wanting information on an airline can change over time. For example, if an airline is considering establishing an off-airport training centre, the local councils, staff organisations and environmentalist organisations of the potential locations will be interested, but the interest may reduce once the decision is made and the centre is operating.

It is easy to establish which banks and financial institutions are currently involved in supporting the airline's business and therefore are interested in the airline's financial position because there is already direct communication; it is more difficult to establish what information potential financiers would like to know. Frequently, a loan agreement will list the information a company must supply regularly to the lender. The Finance Department is probably best qualified to decide on the financial information which will interest current and potential financiers, but the decision to lend is not based solely on financial information. Usually the airline's public relations department will be responsible for establishing who the airline's current and potential future stakeholders are, and it will have an understanding of the subjects which interest them. Sometimes an airline, usually a large one, will also have an investors' relations department, typically related to the public relations department. Smaller airlines may wish to use the services of a specialist investor relations company.

Types of reports

Financial results reporting is usually split into two areas,

- external reporting to third parties, usually in the form of the statutory reports included in an annual, or more frequent, report
- internal reporting, which is to the company's staff.

Although the two reporting streams use the same information and are produced on the same basis, the format and amount of detail are often significantly different because the interests of the audience are different. The general rule for all reports is that the format, whether financial or not, should be designed to meet the needs of the reader(s), and these needs must be kept under review because they change. The external financial reporting will be through the statutory reports where the content and format are described in financial reporting standards, and by stock exchange regulations if the company's shares are quoted by a stock exchange. Although the information included in the internal financial reports is basically the same as in the external reports, there is more flexibility in the format of the internal reports and they usually contain more detail. Even within the range of internal financial reports the same information will be shown in different formats, depending on the use the information will be put to. Tables later in this chapter give examples of this.

Statutory reports

The company's financial statutory reports include:

- A 'Profit and Loss Account', which details and calculates the financial result for the period.
- A 'financial position' or 'balance sheet,' which was discussed in Chapter 5.
- A 'statement of cash-flows', which was discussed in Chapter 6.

These reports contain all the basic financial information which is made available to stakeholders and it is the same information which is the basis for the company's internal financial reporting.

In addition to the reports, there are notes to the reports which include detail for various items in the company's Profit and Loss Account, financial position and statement of cash-flows. These include a 'statement of changes in equity', detailing the changes to the amount due to shareholders e.g., profits not yet distributed as dividends. When the reports are issued in printed form they will generally be in one document together with,

- a report from the Chairman
- a business review, which supplements and supports the financial figures

- commentaries from various directors explaining important aspects of the results together with a description of current and possible future trading conditions.

There is a trend for companies to split the report into sections when including the reports on their website so readers can more quickly access the information they are interested in.

Reporting is a very important function for any company because each member of the stakeholders' group is likely to be interested in different aspects of the operation of the business. Although all will probably want to see the financial reports, some will focus on the general commercial business, others on environmental subjects and others still on future plans.

- Shareholders generally want to be able to assess the value of their shareholding in the company and establish whether it has increased. They also want some return (dividend) on the money they have invested.
- Financiers need to review the information to be confident the company will remain solvent for the foreseeable future.
- Suppliers want to ensure they are providing goods and services to a viable operation and get some idea of the company's plans for the future.
- The staff of the airline are an essential audience. The airline's staff are the people who will actually achieve the airline's long-term strategy. Staff need to know what progress has been made towards the strategic goals. Staff also invest their skills and, in some cases, their career in the airline, and they need to assure themselves that their future is secure. A company's staff are possibly the only people who see both the external and internal financial reports, and it is important that staff can see how the two formats are tied together and use the same figures, otherwise confidence in the figures may be lost. There can be some difficulties in reporting to staff, particularly when the company's shares are listed on a stock exchange and these are discussed later under internal financial reports.

Reporting standards

Guidance on the basis for calculating and presenting the financial figures for the statutory reports is given in a series of 'International Financial Reporting Standards' often referred to as 'IFRS'. These are reporting standards which are designed to be adopted worldwide. Frequently, the IFRS are issued internationally and adopted by a country's accounting institute, in which case they will have a slightly different name but will be essentially the same as IFRS. The IFRS, once part of the nation's reporting standards, will be adopted by the company and appear as part of its accounting policies and practices.

The standards have been issued over a period of years and new or revised standards are issued periodically. Although the IFRS seek to standardise the way financial figures are calculated and reported, there is still room for judgement in the way the IFRS are used. Hence, there are differences in accounting policies between airlines. An example of this is the Depreciation charge for aircraft. Depreciation is a significant cost and can be 10% of total operating costs, hence it is important that the figure is as accurate as is possible. IFRS requires that the Depreciation charge of assets is calculated so that it allocates the cost of an asset, less any estimated sale or scrap value, over the asset's estimated useful life to the airline. The directors are responsible for estimating the variables:

- The life of the aircraft to the airline; this is not the aircraft's total useful life before it is scrapped, but how long the airline plans to use it.
- The residual value, which is the value when the aircraft leaves the airline's active fleet and may be either the sale value or value as scrap.
- The most appropriate way to spread the value of the aircraft (cost less residual value) over the period of its life in the airline's fleet. Most airlines spread the value evenly over the useful life, but there are other options, e.g., spread the value on the basis of aircraft cycles (take-offs and landings).

Bearing in mind how durable aircraft are, it is very difficult to estimate these variables with precision, and for this reason the assumptions will be reviewed regularly and frequently. The annual Depreciation charge is a best estimate which can and is changed periodically. Hence, different airlines using the same aircraft type may have different annual Depreciation charges. Similar problems arise with other charges (e.g., certain aircraft maintenance costs which arise every four or five years), and also estimating which debtors will not pay the amount they owe. These sorts of differences make it very difficult to compare the financial results of airlines, even with similar fleets, despite the introduction of IFRS.

Consistency

An important factor is that the accounting standards must be used consistently, and estimates made systematically, so that for an individual airline, one period's results are comparable with those for previous periods and hopefully for the following periods. If for any reason an important change in an accounting policy is made, the reason for the change and its effect must be clearly described in the statutory reports. The comparative figures for previous periods will be re-stated using the new policy. This is to ensure the results can be compared. The same accounting policies are used when preparing the airline's statutory reports and the internal reports for staff.

External and internal reporting use the same accounting policies and report the same total figures, but the format of the reports is usually quite different. Similar industries within the same country tend to use a fairly similar format for their external, or statutory, reports. The financial reports are supplemented by notes on the detail of some of the figures, and all the figures are compared with figures calculated on the same basis for the previous period. Actually, there is no restriction on the number of periods of figures that can be included for comparison; usually there is a tabulation of the key figures for the previous ten years at the back of the statutory report. In addition to the financial figures, there is a report on the business from the management and a summary of historic figures. The report on the business deals with important information which is not found in the financial figures, it also puts the results into the context of the industry and markets the airline is operating in. This approach helps the reader understand the financial figures and gives information which the reader can use to examine the reports. It would be extremely difficult to produce just one report which meets the needs of everyone who is interested in the airline's financial results and financial position.

Presentation of information

Earlier in this chapter, the probable and different information needs of groups of stakeholders were listed. Although most will be interested in the financial result for the period and the financial position at the end of the period, there will be some information which is only of interest to a smaller group. One example of the differences is amongst investors; this single classification will include,

- long-term investors, who are probably interested in trends and the airline's long-term prospects
- active investment managers, interested in forming an opinion on whether to buy or sell the airline's shares in the short-term
- individual investors, who are likely to have a variety of individual objectives.

The information in the statutory reports together with the supporting detail, commentaries and summary, along with the fairly standardized format, make it possible for each reader to get the financial information they need. The development of statutory reports to include information in addition to the financial information is discussed in Chapter 9. A typical format of the statutory report of a period's financial result is shown in Table 8.1.

The format of the internal financial reports is usually quite different from the statutory reports and is directed to helping the airline's management and staff understand, not only what the financial result for the period was, but also the progress that has been made towards meeting the airline's strategic goals. This is often done by comparing the financial result, financial position and KPIs with the airline's agreed plan for the period. Comparison may also be made with the latest forecast for the following period.

Internal reports are usually a pyramid of reports with summary figures for the Board of Directors, cascading down through the organisation with increasing detail to the sections within the airline's departments. Some reports show

Table 8.1 Profit and Loss Account

	Year 2
	USD
REVENUE	
Passenger Services	14,674,729
Cargo Services	224,719
TOTAL REVENUE	14,899,448
EXPENSES	
Staff	3,442,180
Fuel	2,629,375
Inflight services	525,875
Route	1,367,276
Aircraft maintenance	2,921,378
Aircraft Depreciation	927,803
Aircraft operating leases	123,953
General Depreciation and Amortisation	126,210
Commissions and incentives	21,035
Administration	410,350
TOTAL OPERATING COSTS	12,295,429
OPERATING PROFIT	2,404,019
Finance charges	356,937
Finance Income	14,725
Net finance charges	342,212
PROFIT BEFORE TAX	2,061,807
Tax	70,308
PROFIT FOR THE PERIOD	1,991,499

Note: The figures for the previous year would be shown in a column beside Year 2.

the same information in a different format to highlight different aspects of the operation, for example, an indication of the financial profit or loss from each route. These are discussed later in this chapter. A major difference between the external and internal reports are that, whereas in the external reports the financial result is compared to the previous year's figures, in the internal reports, the result is compared to the airline's agreed plan for the period. In addition, figures are shown in more detail to give managers and staff a view of how the airline is performing financially, and there will possibly be figures showing trends as well.

Table 8.1 shows a frequently used format for reporting the calculation of the profit or loss for the period using the normal statutory report format. The cost figures include some estimates, like the Depreciation charge. To help the readers of the statutory report understand the numbers and to help make comparison with other companies, a variety of financial 'results' other than profit or loss have been developed. The most frequently used and referred to are:

- EBIT, means Earnings before Interest and Tax
- EBITDA, means Earnings before Interest, Tax, Depreciation and Amortisation
- EBITDAR, means Earnings before Interest, Tax, Depreciation and Restructuring or Rent cost.

The reason some legitimate charges are excluded from the calculation of the financial result are,

- Interest: this charge depends on the amount of borrowings; some airlines have a high level of borrowings compared to others who finance their assets with equity.
- Tax: the rate of taxes on profits varies significantly between countries, as does the basis of calculation. Also tax laws and rates change frequently.
- Depreciation: this is a substantial charge which is a 'best estimate'.
- Amortisation: similar to Depreciation, this charge is a 'best estimate'.
- Restructuring: this is the cost of making significant changes to an airline's organisation and, as such, are a one-time cost, hence they will not re-occur.
- Rent: similar to the argument in Depreciation, some airlines choose to rent assets (e.g., aircraft hangars and offices), while others choose to fund them through loans or equity.

The different calculations for the financial result seek to eliminate differences in the way airlines are organised financially, so their results can be compared. For example, some airlines may be financed mainly by money from shareholders. Hence, they have a low interest charge, but pay more in dividends, while others may be financed primarily by loans, which therefore have a higher interest charge. These approaches are shown in Table 8.2.

Table 8.2 Profit for the period

	EBIT	EBITDA	EBITDAR
	1,991,499	1,991,499	1,991,499
Add back;			
Net finance charges	342,212	342 212	342,212
Tax	70,308	70 308	70,308
Aircraft Depreciation		927 803	927,803
General Depreciation and Amortisation		126 210	126,210
Aircraft operating leases			123,953
EBIT	2,404,019		
EBITDA		3,458,032	
EBITDAR			3,581,985

Whilst it is accepted that these presentations attempt to provide a consistent basis for calculating a financial result which can be easily compared to the financial results for previous periods, as well as with the financial results for other companies in the same or similar business, it must be remembered that each calculation excludes important costs which nevertheless have to be paid and charged against the revenue earned. It cannot be assumed that any of these presentations form the actual final financial result for a period.

Internal financial reports

These reports include all the reports used by the airline's management to manage the company's operation, as well as the communication of the airline's financial result and position to the airline's staff. Both of these types of report are likely to contain information which is confidential to the business. Nevertheless, managers need good quality financial information in order to do their job. One way to protect confidential information is to ensure that each manager gets all the financial information they need, but not any additional financial information which is confidential. During the process of appointing a manager, some assessment will be made of their ability to protect the company's secrets.

Care is also needed in deciding what financial information is communicated to staff and when. Clearly, the financial information in the airline's statutory reports can be communicated because it is available to anyone who asks for it, but it should not be released to staff until it is published generally.

Dealing with financial results, particularly forecasts, is a problem for a company whose shares are quoted on a stock exchange. Generally, the stock exchange's rules will include a requirement that any 'price sensitive' information must be announced to the public as soon as it is available. Telling the airline's staff the forecast financial result or financial position may be considered the equivalent of making an announcement which should be made to the public. This is the same problem faced when deciding how to explain the airline's strategy to its staff and the solution will probably be similar; the Board of Directors has to find a way to communicate the plans to staff in a way which does not breach the stock exchange's rules. Some companies only give a summary of the forecasted financial information to staff, while others communicate only 'forecast operational information'.

At this point, it is worth remembering that the objective of reporting the detail of the financial results to managers is so that they can make comparisons with trends, past performance and the company's plan, in order to identify potential future problems and learn from past performance, thereby making better decisions in the future.

One of the differences in presentation between external and internal financial results reporting is that the statutory reports will generally report the information in total amounts, classifying revenue and expenses by their nature and generally use headings in the financial position and cash-flow statements.

Internal results report the same information, but analyse revenue, expenses, assets and liabilities in detail and in various ways to give different insights into the operations of the company, as well as to help managers control the revenue, expenses and assets used in the business. Examples of some of these reports are shown in Tables 8.3, 8.4 and 8.5, but in practice every airline will have its own preferred way of analysing and reporting the detail of its operations. Generally, managers will only receive the level of detail they need to manage (i.e., directors will probably see the whole financial picture in a summary, while a section supervisor will see the detail of the revenue and expenses they control and influence).

- Responsibility reporting. In this report, all the items making up the revenue and expenses are split into the areas under the responsibility of various executive directors. The assumption is that some expenses can be identified to individual departments (e.g., staff salaries), but other expenses

Table 8.3 Responsibility report

| | | Year 2 | |
| | | USD | |
Responsible manager	Heading	Actual	Variance from plan
Passenger sales	Revenue – passenger	14,674,729	
	Commission and incentives	21,035	
	Staff	494,490	
	Staff – station offices	53,810	
Cargo sales	Revenue – cargo	224,719	
	Staff	74,830	
Inflight services	Inflight	525,875	
	Staff	224,490	
Routes	Route	971,361	
	Landing and parking	395,916	
	Staff	129,660	
	Tax	1,420	
Finance	General Depreciation	126,210	
	Finance charges	356,937	
	Finance income	14,725	
	Tax	68,888	
	Staff	201,370	

And so on through each responsibility.

Note: the variance from the plan will be the figure for action.

Table 8.4 Operating profit

	USD	Basis	USD			
			Route result			
			1	2	3	4
REVENUE						
Passenger	14,674,729	Actual	2,201 209	2,934 946	5,136 155	4,402,419
Cargo	224,719	Actual		137 016	87 703	
- Commission	21,035	Actual	517	2 714	9 493	8,311
TOTAL	14,899,448		2,200 692	3,069 248	5,214 365	4,394,10 8
EXPENSES						
Direct costs	5,796,057	Actual	813 595	1,147 426	2,045 098	1,789,938
Indirect costs						
Aircraft maintenance	2,921,378	Block hours	438 207	584 276	1,022 482	876,413
Aircraft Depreciation	927,803	Aircraft value	71 369	71 369	356 847	428,218
Aircraft operating leases	123,953	Flights	16 527	24 792	41 317	41,317
General Depreciation and Amortisation	126,210	Passenger revenue	18 932	25 242	44 173	37,863
Administration costs	2,580,413	Passengers	387 282	527 172	948 367	717,592
Net finance costs	342,212	Contribution	52 264	72 412	119 414	98,122
Tax	68,888	Contribution	10 521	14 577	24 038	19,752
TOTAL			1,808 697	2,467 266	4,601 736	4,009,215
PROFIT	1,991,499		391 995	601 982	612 629	384,893
% of revenue	13.4		17.8	19.6	11.7	8.7

Table 8.5 Contribution report

Year 2

	Total USD	Route result USD			
		Route 1	Route 2	Route 3	Route 4
REVENUE					
Passenger	14,674,729	2,201,209	2,934,946	5,136,155	4,402,419
Cargo	224,719		137,016	87,703	8,311
– Commissions	21,035	517	2,714	9,493	
NET REVENUE	14,878,413	2,200,692	3,069,248	5,214,365	4,394,108
DIRECT COSTS					
Cockpit staff	848,300	127,645	169,260	297,905	253,490
Cabin staff	423,810	63,972	87,361	148,334	124,143
Inflight	525,875	68,881	95,175	204,056	157,763
Route	971,361	132,704	195,272	338,976	304,409
Landing and parking	395,916	21,945	59,183	147,566	167,222
Fuel	2,629,375	398,416	540,895	907,251	782,813
Tax	1,420	32	280	1,010	98
TOTAL DIRECT COSTS	5,796,057	813,595	1,147,426	2,045,098	1,789,938
CONTRIBUTION	9,082,356	1,387,097	1,921,822	3,169,267	2,604,170
% of revenue		63.0	62.6	60.8	59.3
Administration	7,090,857				
PROFIT	1,991,499				

Note: The contribution may be measured against other factors, e.g., per block hour, per passenger, if that helps management.

covering the whole company are decided by one person and then charged to departments. For example, the Property Manager would probably negotiate the rent of office space on behalf of the company, but the rent expense would be charged to each department on some basis, probably on cost per square foot of space. In Table 8.3, the CFO is shown as responsible for the company's Depreciation charge on all equipment other than aircraft, but the total charge of USD 126,210 would be allocated to each of the departments using the equipment. The responsibility report will be supported by reports analysing the figures into the responsibilities of the individual managers reporting to the executive director to be used in the control of both revenue and expenses. There should also be a similar report in which the items in the airline's balance sheet items, particularly the working capital items, are similarly analysed into each director's responsibility (see Table 8.3).

- Route operating result. This report takes the figures in the total Profit and Loss Account and calculates the result of each route, profit or loss, which when added together equals the airline's profit. The report may be supported by details of the revenue and expenses of individual flights or routes showing the profit arising from individual operations. The cost of support services (e.g., Planning Department, Finance Department) are apportioned to each route using some agreed basis which gives an indication of how much of the support services the route uses. This approach which seeks to indicate the result of each route or operation is often referred to as being 'fully costed', because all of the airlines support expenses are allocated to operating routes (see Table 8.4).

- Contribution from route operations. This is a different approach that allocates all revenue and expenses which can be clearly identified with a particular route to that route, but does not apportion the cost of support services to each route on an arbitrary basis. It acknowledges that it is virtually impossible to accurately identify the support and administration expenses with a particular route. All the support and administration expenses are treated as one pool of expenses, and each individual flight or route makes a contribution towards these general fixed costs. Once those costs have been paid, the amount remaining is the profit. The benefit of this approach is that it gives a more accurate basis for ranking the importance to the airline's financial result of each route, both as a percentage of revenue and in total money. Looking at the route contribution also gives an indication of the possible effect of changing the frequency or cancelling an individual route, but before a decision is made to change each item, it would have to be examined to assess individually what effect a change would have. For example, cancelling a route may affect the number of passengers connecting to another of the airline's routes. Not all of the route costs would disappear unless aircraft were sold, staff dismissed, etc. Although this presentation only gives an indication of the result it can be the trigger for a detailed review of the viability of a particular route. Calculating and measuring the contribution from each operation will help management focus on areas where improvements are needed (see Table 8.5).

These three approaches above use a different basis for analysing revenue and costs; 'allocation' and 'apportionment'. Allocation is where the whole of a revenue or expense item can be identified with an individual manager (responsibility) or individual operation (contribution). For example, the revenue from inflight sales can be identified with an individual flight, and stationery expenses can be identified with a department or section. Apportioning seeks to spread all non-operating revenue and expenses items to the airline's routes on some acceptable basis. It is probably easiest to understand these approaches through examples.

- When calculating the operating cost of each fight or route in Table 8.4, the cost of running the Finance Department, which is the CFO's responsibility, will be apportioned to each flight or route on some agreed basis; for example, the number of passengers. When calculating the contribution from each flight or route in Table 8.5, the total cost of the Finance Department will be shown in the total of Administration, sometimes referred to as 'Overheads'. This shows that the Finance Department is an overhead and a 'fixed cost' (i.e., one that is generally unchanged by the level of operating activity).
- The cost of inflight meals will be allocated to the responsibility of the Inflight Services Director (Table 8.3) and allocated to each individual flight or route in the other reports (Tables 8.4 and 8.5). This shows that the expense is a 'variable cost' (i.e., one that varies with the level of operating activity).
- The total cost of cockpit crew will be allocated to the responsibility of the Flight Operations Director as in Table 8.3. The cost of cockpit crew when flying will be allocated to individual flights or routes as in Table 8.4, but when on training or on standby, the cost will be an overhead and apportioned to flights and routes on some agreed basis. The same principle applies to cabin crew. Therefore, cockpit and cabin crew costs are not 100% fixed nor 100% variable, but will generally be classed as variable or semi-variable depending on how sophisticated the airline's systems is.

Accrual accounting

Revenue and expenses are recorded in the airline's accounts when the transaction is agreed, not when money is received or paid. Revenue is recorded in the Profit and Loss Account when a passenger travels or cargo is carried, and expenses are recorded when a third party provides goods or services. If the cost of goods or services is not known, an estimate will be recorded and adjusted later. This approach is called 'accrual accounting', and is often described in the airline's statutory reports in a Statement of Principal Accounting Policies. This approach also ensures that the revenue for a period is matched with the costs needed to produce it. Hence the result, revenue less expenses, is as accurate as it is possible to calculate, and follows the principle of matching revenue with costs.

If the revenue has been earned, i.e., the passenger has travelled, but the passenger has not paid the fare, the amount due is shown in the airline's balance sheet as a 'debtor'. Similarly, if the supplier has provided some goods or service but not yet sent an invoice or the invoice has not been paid, it is recorded in the balance sheet as a 'creditor'.

Classification of expense accounts

All expense items are either charged against revenue or 'capitalised'. Hence, these are usually referred to as either 'revenue costs' or 'capital costs'. Although most of the expenses of an airline are related to its day-to-day operations, some of the money spent will be for items which the airline will use and get a benefit from for many years. It would not be right or fair to treat these purchases as an expense for just one period. Examples of expenses which have a long-term benefit are, the cost of:

- Aircraft
- Vehicles
- Computer programmes
- Replacing or renovating the passenger cabins of aircraft
- Re-modelling offices

These expenses are treated as assets and included in the 'non-current assets' section of the balance sheet. The cost of tangible assets, like aircraft, are charged to the results over their assumed useful life after taking into account their assumed sale or scrap value. Intangible assets are treated in the same way, but generally the period is shorter and there is not an assumed sale or scrap value.

When an expense is treated as an asset, whether tangible or intangible, it is said to be 'capitalised', which means it is added to the long-term assets of a company. It is important to establish rules for what expenses can or cannot be capitalised to ensure the asset figure is as accurate as possible and the financial result for the period includes all the costs incurred for that period. The rules can be very detailed, particularly for something as complicated as the purchase of an aircraft. They need to cover many items, such as the cost of a representative at the manufacturer, interest on any loan used for pre-delivery instalments, the cost of the delivery flight and any other costs which can be identified solely with an aircraft. The cost of the aircraft will not, generally, include spares, inventory or flight simulators. The list is substantial. The general principles guiding what expenses may be capitalised are as follows:

- The asset is expected to produce benefits for the company over more than one reporting period.
- All benefits must be quantified or defined.
- The cost of the asset must be charged against the result of future periods in a systematic way to write off its cost over its useful life to the airline. In the case of a tangible asset this charge is called Depreciation and for an intangible asset it is called 'Amortisation'.
- Any additions to an existing asset should be expected to increase its useful life or increase its benefit to the company.
- The purchase of every substantial asset and any significant additions to an asset must be approved in advance by the company, usually by the directors, but in some cases by a sub-committee of the Board of Directors.

There are some problematic expenses, such as major advertising campaigns, where it is difficult to decide whether the expense should be capitalised or immediately charged against revenue. In the case of an advertising campaign, the expenses are incurred for good reason, but it may be difficult to quantify or define the expected benefits or predict how long any benefits will last. Where there is any doubt it is usual to treat the expenses as a cost.

The capitalised items appear in the non-current assets section of the balance sheet and are split between tangible and intangible assets. A note in the accounts shows the changes to the amount shown as non-current assets in the balance sheet and may appear as:

- Additions
- Sales or the value of items scrapped
- The amount charged as Depreciation or Amortisation.

Financial reporting can be split into two types: statutory reports and management reports. The format of statutory reports is fairly standardized, follows industry practice, is governed by national and international reporting standards and these standards are frequently reviewed and updated. The figures reported tend to be total figures.

The format of management reports is more flexible, and is designed to provide the information the airline's management need to effectively manage its operations and includes more detail and analysis. In both cases, the basis for calculating the financial information uses the same set accounting principles.

The airline's financial information can be thought of as a container full of information, which can be presented in many different ways, but will always produce the same financial result.

9 Future reporting

Change is constant

For very many years the main focus of statutory reporting has been on the company's financial result and financial position, and trying to ensure this information is as accurate and well-explained as possible. This concentration of effort naturally led the management of companies to also concentrate on the company's financial result and financial position, in addition to monitoring the company's operations. Each of the changes to the statutory reporting requirements has had a strong influence on the formats and techniques used by companies in their own internal financial reporting for management.

Originally, society's view of a company was that it was just a source of profits, and therefore dividends, to shareholders. In addition, it was an employer of people. This view developed and changed when more individuals became interested in owning shares in companies. There was the need to explain the company's financial result and financial position in more detail, but in a way that could be understood by individuals regardless of their level of financial knowledge. In recent years, society's view of the role of a company has developed into viewing each company as a part of society, using available resources, both physical and intellectual, and having an effect on society in general. This change in attitude is having a significant influence on what companies report to society (i.e., their stakeholders) and, in turn, this is influencing the internal management reporting in companies. As an indication of the changes to statutory reporting, 10 years ago, very few (if any) airlines included any information on their 'GHG emissions' (i.e., greenhouse gas emissions) in their statutory report, their use of utilities (e.g., electricity), their participation in charitable organisations, or their support to those of their staff who wished to do charitable work. It is likely that the reporting requirements for companies will continue to change and, as a consequence, each company's strategy and its definition of 'success' will be changed and expanded. The changes will influence the company's internal reporting.

The amendments to reporting need to be put into the context of a company's reporting system, both financial and operating. The objective of a company's accounting systems is to collect data (i.e., facts) and present the information in a format that helps managers run their business, enabling them to clearly and accurately report on their current financial achievements to shareholders and other interested parties. The formats and techniques used in statutory and internal management reporting have been developed and changed over years to improve the usefulness of financial reporting to management.

Change will continue

Although International Financial Reporting Standards have been established for many years, changes to them are still being made, either to refine a standard as a result of experience gained by companies using it or to deal with another topic. Attempts at defining a company's relationship with the rest of society and its environment have been in progress for many years, and although the process has started, it has not finished. In many jurisdictions public companies are starting to report on non-financial matters. The development and refinement of a company's responsibilities, its place in society and effect on its environment has resulted in, and been accompanied by, the inclusion of more financial and operating information in a company's statutory report to help the reader understand the company, how it works, what it is doing and what the results have been. These changes have rolled on into the company's internal reporting systems in the form of additional records and measurements together with additional reporting, some of it financial and some non-financial. This process has not finished and more changes are to be expected in the future, not the least of the changes will be expanding the current non-financial reporting to non-public companies.

The current reporting requirements for companies are part of a trend which will change statutory reports from the current approach of making them a mainly a financial report with business commentaries to a more comprehensive report which describes for every company:

- The current condition of the business as a whole; not just its financial aspects, but also the commercial position, including the director's assessment of the company's opportunities and the risks it faces.
- How the company is run and how it plans to achieve its objectives.
- What resources the company uses, both materials, e.g., water, and community, e.g., people and their skills, and how well the company uses them.
- The effects of the operation of the business on the community, both locally and more generally.
- How the company supports and helps the community.
- The financial result and financial position.

The International Financial Reporting Standards are part of this trend and process. It can be expected that at some stage there may be a requirement to include an estimate of the financial effect to each piece of information in the environment and social parts of the report (e.g., the value or cost of helping staff to participate in charitable works).

Although on the surface it appears that the changes to statutory reporting are being initiated by national governments, regulatory organisations like stock exchanges and professional bodies such as accountants or investment analysts, the real pressure is from the society in general in the economically developed nations. The changes are happening gradually, but quite steadily. The organisations and regulators involved in the process of guiding the developments and changes are required to discuss each proposed development or change widely with those it is likely to affect before it is introduced. It is sometimes difficult to reconcile the differing opinions on what is involved in each change and how workable the change is. A very difficult part of the process of developing each topic or subject is to produce a clear definition of the objective of the requirement, why it will be beneficial and who will benefit, in addition to how the requirement should be introduced. The organisation championing the change also needs to take into account that most, if not all, new requirement or modification of an existing requirement will probably mean additional work, hence cost, to the companies affected.

For changes to financial reporting, it is usual to issue a draft of the new reporting standard with a covering explanation of the reasons for the change and the expected outcome together with an invitation for companies and professional organisations to comment on the detail in the draft. The draft standard is usually called an 'exposure draft'. Naturally there will be a time limit before which companies and organisations should make their comments and observations, but the permitted period for replies is often quite long. This is because once it has been approved, the proposed standard will apply to every type of business in all industries, every type of business organisation and all companies, whether they be small or large.

There are currently three main themes which are being progressed in the area for reporting:

- Continuing with the development and refining of financial reporting standards; new standards and exposure drafts are issued frequently.
- Reporting on the company's business model (i.e., how the business will achieve its long-term objectives); this will involve defining the business the company is in, who the target customers are, what factors will differentiate the company from its competitors, what its opportunities are and what risks it faces.
- 'ESG'. These initials stand for Environment, Social and Governance; ESG is an overarching term that covers important information which over the long term is likely to impact a company's performance and financial value. It includes the company's effect on the environment, on society and also describes how the company is run.

These themes are being developed, refined and discussed with the business community by different organisations. The organisations co-ordinate their work to ensure that all of the guidance produced works together to produce a consistent theme, and that the timing for the issue of new standards or guidance does not conflict. Nevertheless, to many businesses, it seems there is a continuous stream of required changes.

A number of international and national organisations have produced frameworks and guidelines to help companies determine what they should report on. In addition, they have developed definitions for each of the items to be reported which will help to make reporting consistent. The lead organisations developing and refining the major initiatives are:

- The Integrated Reporting Council, which helps companies report concisely on how they create value over time by explaining how a company's business model changes the value of the financial, social and natural capital it owns or controls. This approach is really looking at the company through the eyes of its directors. It is difficult to give a precise definition of social capital and natural capital as they are difficult to measure, but it is the council's role to solve this difficulty.
- International Financial Reporting Standards are developed by the International Accounting Standards Board for reporting financial items. Many countries adopt these standards and issue them as their own national reporting standards.
- The Global Reporting Initiative develops guidelines for reporting on Environment, Social and Governance matters. These guidelines provide a process for determining what topics are important for a company and gives data definitions for these items. The GRI guidance permits three different levels at which companies can follow this standard with the thought that companies will start at the simplest level and eventually progress to the most advanced level.
- The Sustainability Accounting Standards Board has developed reporting guidance for many major business sectors, assessing and defining which issues are important for that sector. The thrust of this work is to have consistent definitions used within each industry.

In general, it is not mandatory for all companies to follow the standards and guidance issued by these bodies, but in many jurisdictions, all, or extracts from, the standards and guidelines have been adopted by the national accounting standards institute and/or by regulatory bodies (such as a stock exchange), and they are issued as national reporting requirements and guidelines. For example:

- International Financial Reporting Standards will be adopted by national accounting regulators and issued as national financial reporting standards.
- ESG standards and guidelines, entirely or in part, may be adopted by a stock exchange's regulators and become required reporting for companies listed on the stock exchange, regardless of the nationality of the company.
- Extracts from the advice issued by the Integrated Reporting Council on the description and explanation of a company's business model may be

incorporated into a country's company law or issued as guidance by the stock exchange regulators.

The approach of developing standards and guidelines which can be used internationally means that national standard setters and regulators can quickly examine and adopt the international standard as their own. This is having a significant effect in standardising reporting, both financial and non-financial, of the developed economies and provides a wide range of experience of using the standards and guidelines, which is useful when considering whether any standard or guideline needs to be revised.

Size does not matter

These developments are not solely for large companies or those whose shares are quoted on stock exchanges, although many of these companies have assisted in developing and trialling the standards before they are issued. The developments are also for small and medium-sized enterprises, and this would include a start-up airline, often described as 'SMEs' and sometimes 'private enterprises. Some would argue that SMEs can gain greatly by adopting the financial reporting standards and other guidance as early as possible in the growth of the company. It would help the management of the SME to have a clear idea of how their business model works. Improved reporting to their shareholders would also improve the information given to staff and, importantly, the company's financiers and any potential shareholders and financiers.

If an SME would rather not follow the full reporting standards there is often a national reporting standard and/or guideline for SMEs, it may be called something like the Financial Reporting Guidance for SMEs. There may also be advice on how an SME's financial and operating reporting should develop as its business grows. One problem an SME may experience is finding the resources and developing the systems required to provide the information needed, but there may be help and guidance in this area as well. Another problem often encountered by SMEs is trying to decide just what should be reported. There is guidance in each of the standards and guidelines, but there is also the general guidance that the SME can fall back on, which is to 'only report what is important, rather than every minute detail'.

When new financial reporting standards are first introduced, they are not usually mandatory immediately. Usually companies will have one- or two-years leeway before having to follow the standard. However, for more general reporting guidelines, there may be one or two stages before the standard becomes a requirement after a few years. The first step may be for it to be entirely voluntary, whether or not a company follows the guidelines. This period may then be followed by a time during which companies not meeting the guidelines must explain in their statutory reports their reasons for not complying. After a while, sometimes years, the approach will probably change to requiring that the guidelines are to be followed. Eventually, once the standards and guidance have been proven and tested, they will become mandatory. This apparently slow approach is appropriate because of the wide variety of businesses involved, the need to ensure that information which is confidential to the business is dealt with correctly, and to take into account the changing requirements of society.

Naturally, every company will require systems to collect the information needed to be included in their reporting. In addition, staff must also be trained in exactly what information is needed. It takes time and money to develop and prove systems and to complete staff training. Similarly, the company's internal reporting systems should be augmented to permit budgets to be developed for the new reporting items, if that is appropriate, and to report the actual results of non-financial matters in a similar way to reporting financial results.

Frequently, the Finance Department takes the lead in developing the systems and designing the statutory and internal reports, but this does not mean it is purely a financial matter. The objective of these reporting developments is to expand the reports from a primarily financial report to a description of the company as a whole, by supplementing and supporting the financial reports. Hence, all departments become involved in the contribution.

The social area of company reporting will continue to develop because solely financial reporting does not give a complete picture of the company's operation to all those who are interested in the company. There will be continuing pressure to expand the scope of company reports while ensuring they are easy to read and understand. In some jurisdictions, it is permitted for information which does not change frequently, such as the way in which the Board of Directors and its sub-committees work, as well as their responsibilities, to be detailed on the company's website and simply referred to in their statutory reports. Some companies have split their statutory reports into separate short sections which focus on just one aspect of the business's operation, e.g., 'Environment', 'Strategy' and 'Finance'. This approach has the advantage that the reader does not have to read the whole report to get the information they want. All of these developments make it clear that statutory reporting is a communication exercise.

For all

These advances are not just for the external readers of the statutory reports, but for the company's staff as well. The company's strategy may have to be amended to include long-term objectives for the environment and the social effects the company has. The company's staff are a vital audience because they are the people who actually achieve the company's strategy and they must understand all the elements of it and what progress has been made. Reporting changes are being driven by pressure from society and company staff are part of that society.

Company reporting will continue to change and improve for many years as experience is gained and readers' reactions are understood and analysed.

Glossary

These are the definitions used for terms in this book. In other circumstances and books a different definition may apply.

Accounting   the process of correctly recording regularly and systematically in the airline's 'books of account' or records, all its business transactions that result in the transfer of money or money's worth, so the reader knows the true state of the business.

Accounting policies   a list of principles and procedures governing the accounting and books of account agreed by a company's directors and management, which is used when preparing the company's statutory and management financial statements. Accounting policies are not the same as accounting standards but are the way the company has interpreted accounting standards or decided to deal with an accounting matter not covered by an accounting standard.

Accrual   a method of accounting that records an economic event when it is performed, regardless of when any payment due is made. For example, the value of equipment purchased is recorded when it is received, regardless of any credit period which permits payment to be made later.

Active investment manager   this term describes a manager of a portfolio with the strategy of making individually selected investments which they believe will outperform an investment benchmark index or a target return.

Aircraft utilisation   an efficiency measure for the amount of time an aircraft is operating compared to the time it is available for operations. The measure is often expressed in hours.

All-in cost of debt   the all-inclusive cost involved in a borrowing transaction. Included will be all interest payable, establishment fees for bank accounts, early termination costs, routine third-party administration costs during the life of the borrowing (e.g., the administration of a trust).

Allocate   in the context of accounting, the term means to charge to a route or other profit centre only the revenue and expenses which can be solely identified with that route or profit centre. The term also has other meanings associated to it.

Amortize to charge against profits the cost of an intangible asset over the period it is assumed to be of value.

Apportion to divide and charge revenue and expenses items in accordance with using a pre-determined basis or rule which is used consistently for more than one period.

Assets – 'free' assets An asset is an item that a company owns which it believes will provide future benefits in excess of its cost. The cost of the asset appears in the company's balance sheet. Assets can be used as security for loans made to a company. If an asset has not been used as security it is called a 'free' asset.

Audit an independent, regular review of a company's financial accounts, using as many of the company's documents as are relevant, in order to establish whether the accounts accurately show the company's financial performance and financial position.

Audit committee a sub-committee of the Board of Directors responsible for overseeing and advising on the content and standards of the company's financial reporting. If the company's shares are quoted on a Stock Exchange, the minimum duties of this committee may be specified in the Stock Exchange's rules. The Board of Directors may add to these responsibilities.

Available seat miles a statement of an airline's passenger-carrying capacity. The calculation is the number of seats available over the airline's fleet multiplied by the number of miles flown. Some airlines prefer to use kilometres rather than miles, in which case the measure becomes 'available seat kilometres'.

Avoidance of Double Taxation Agreement an agreement between two countries detailing how to avoid the levying of tax in both countries on the same declared profit. Frequently, the approach is for each country to mutually exempt from taxation the profits earned in one country of companies registered in the other country.

Balance sheet (Statement of Financial Position): a financial statement that reports the detail of a company's assets, liabilities and shareholders' equity at a specific time. The report is supported by Notes to the Accounts, which include additional detail for some items. The figures are a mixture of historic figures and current valuations; therefore care is needed when reading and using the figures. A balance sheet is often described as a financial statement giving a snapshot of what a company owns and owes.

Balloon loan a loan where regular repayments are made during the period of a loan, but these are not sufficient to repay the total borrowed by the end of the loan; therefore there is a large repayment due at the end of the loan period.

Basis swap an agreement which effectively changes the basis upon which interest is charged on a loan. The agreement involves the exchange of funds calculated on the different basis to leave one party paying and the other party receiving amounts calculated on the different basis.

Benchmark rate a reference interest rate used to set other interest rates, e.g., London Interbank Offered Rate. Usually the rates are published in the financial press; therefore to some extent they are an independent source of information. Using a benchmark rate gives the lender a return which is related to the current market rate market. Similarly, the borrower pays interest at an approximate current market rate.

Best estimate an independent estimate which is assumed to have the best probability of being the most accurate.

Best, most likely, worst descriptions of different types of plans or forecasts. 'Best' is the most optimistic, 'most likely' is the one with the highest chance of being accurate, 'worst' is the position where many things go wrong.

Board of Directors the governing board of a company elected by shareholders at the annual meeting of the shareholders. The Board of Directors are responsible for the future direction of the company and policy-making.

Bond a bond is a loan made by an investor to a borrower, which is evidenced by the borrowing company issuing a certificate. The certificate lists the interest the borrower will pay and when, when the loan will be repaid and whether the loan is secured by any of the borrower's assets.

Bond maturity the date when a company issuing a bond must repay the principal to the holder of the bond. Failure to make the repayment on the maturity date results in the company being 'in default' and any assets allocated as security for the bond may become forfeit.

Bond rating an assessment made by an independent rating company of the probability that a company issuing a named bond will default by not paying interest due on the bond or repaying the principal when due. There are a number of classifications of the probability of default (e.g., AAA to D). Investors are influenced by the rating when considering whether to buy a bond.

Books of Account the financial records, supporting files and analysis that together record all the detail of a company's financial transactions.

Budget a detailed estimate for a period of expected revenue, expenses and resulting profit together with capital transactions which together give an estimated future financial result and financial position.

Bullet loan: a loan that is fully repaid by one lump sum payment at the end of the loan period. Sometimes the bullet payment will include the interest due as well.

Business plan a document prepared by a company's management and then reviewed, amended and approved by the company's Board of Directors, which details the company's long-term business targets, the approach to achieving the targets and a timetable. The plan with have a separate section for each of the major parts of the business and the financial results. It will also probably describe the company's business, industry background and business tactics. In essence, the document describes the company's current view of its future.

Cap (finance)   an agreement with a third party which agrees a ceiling or upper limit on the price of a commodity or a financial transaction (e.g., an interest rate).

Capital cost   the amount paid for an asset which will benefit the company's operation for more than one accounting period, usually for many years (e.g., aircraft, hangars). The benefit may be the direct generation of profits, increasing productivity or the reduction of costs.

Capital structure   the mixture of equity (funds from shareholders), long-term and short-term debt by which a company finances its operations.

Capitalise   to charge the cost of a purchase or an expense to the capital cost of an asset, rather than charging it directly against profits, e.g., bank charges incurred remitting the purchase price of an asset may be included in the cost of that asset.

Cash-flow   money or money equivalents received and paid.

Centralised   locating all the department's staff and operation in one place.

Checks and balances   a system which splits the responsibilities of a company between different departments in order to ensure that decisions are made impartially, and to reduce the possibility of fraud.

Collar (finance)   an agreement with a third party which agrees a ceiling limit and a floor limit on the price of a commodity or a financial transaction. If the price moves outside of the ceiling or floor one party will compensate the other.

Comfortable (risk)   to be relaxed about the adverse effects of a risk (i.e., not having any stress about the matter).

Commercially sensitive   commercial information that, if known and used by competitors, would damage a company's business or commercial dealings. Examples of potentially commercially sensitive information are the ingredients and formulae of a company's product, the profit or loss margins on individual products, routes or discrete business activities.

Company Secretary   an officer of a company responsible for keeping a record of directors' and shareholders' meetings also for the legal records of the company, including a register of shareholders. Other duties may be allocated when decided by the Board of Directors.

Company secrets   a company secret is a company's property, which can be a formula, practice, process, design, commercial technique, strategy or plan that is not generally known within an industry, but if it were known could benefit competitors or customers.

Contra deal   a form of barter agreement where two or more companies exchange goods and services without cash passing between the parties. Although no cash passes the value of the contra deal, it must be recorded in each company's books of account.

Cost of borrowing   the interest and other charges which a company has to pay when it borrows money.

Counterparty risk   the risk that a party to a contract will not perform its obligations under an agreement.

Creditors Policy   a statement approved by the board on how the company's creditors should be managed.

Currency exposures   the risk of a monetary loss because the value of an asset is expressed in a foreign currency and not in the accounting country of the company.

Currency Risk Assessment   the process of establishing the risk to a company of loss (or profit) from transactions not in the company's home currency.

Currency swap   an agreement between two parties by which they agree to exchange currencies. The agreement is usually used to change the currency of a loan.

Damp lease   an agreement by which the owner of an asset, e.g., an aircraft, agrees to lease the asset to a third party, together with some of the personnel required to operate the asset. Generally, for an aircraft this means the owner will not provide cabin crew. There are other names for this type of arrangement and different definitions. The term can be thought of as describing variations of 'wet' or 'dry' leases.

Debt Servicing   the amount of money needed to pay the interest on outstanding loans and bonds together with repayments of principal during a period. If a repayment of principal is not due during the period, an amount can be added to the total interest as an estimate of the amount which would have been repaid had the loan or bond been repayable in equal instalments during its life. If the company has a loan or bond which is not required to be repaid in equal annual instalments, but nevertheless decides to put aside an amount of money to accumulate towards future repayment(s), the accumulated fund is called a 'sinking fund'.

Debt/equity ratio   the ratio between a company's debt, i.e., the money borrowed from third parties such as banks and its equity (i.e., the money shareholders have invested plus the profits which have not been paid out as dividends to shareholders). Equity is sometimes referred to as 'Shareholders' Funds'. The ratio may be called the 'gearing' or the 'gearing ratio':

Debtors Policy   a statement approved by the board on how the company's debtors should be managed.

Decentralised   locating the department's staff and operations in the departments they serve.

Depreciation   Depreciation is an accounting method of allocating the cost of a tangible asset over its useful life and is used to account for declines in value.

Derivatives   a marketable contract whose value is related to an underlying item, this might be an asset like shares, an index like the 'Jet Fuel Price Index', or a financial transaction like an interest rate or foreign currency exchange rate.

Dividends   a distribution by a company to its shareholders of all or part of the company's profit. Generally, not all of the profit will be distributed, and the undistributed portion will either be re-invested in the company's business or held in reserve in case there is a year without profits. Dividends may be paid at any time during the year. Some companies pay a dividend at the end of each half-year or even every quarter. The dividend paid in respect of the company's result for the full financial year is usually called a 'final dividend'. Depending on the law of the jurisdiction in which the company is registered, the final dividend must usually be approved by a meeting of the shareholders.

Double Entry Accounting   the basic assumption supporting a company's accounting, which is that every financial transaction has two aspects, first, to either increase the company's assets or reduce its liabilities, and the other to either reduce the company's assets or increase its liabilities. For example, buying a hammer for cash, reduces a company's Cash (asset), but increases the amounts of its Stocks (asset).

Dry lease   an agreement by which the owner of an asset, e.g., an aircraft, agrees to lease the asset to a third party. In this agreement, the operator provides the personnel required to operate the asset.

ESG reporting   the system of preparing the report and the report on a company's Environmental, Social and Governance matters.

EBIT   these initials stand for 'Earnings Before Interest and Tax'. The figure is an accounting figure that calculates the profit for a period earned from all a company's business operations, excluding interest expenses for the period and taxes on profits. The figure is also referred to as 'operating profit'.

EBITDA   these initials stand for Earnings Before Interest, Taxes, Depreciation and Amortisation, and is an alternative measure of a company's financial performance. Its main value is as a comparison between companies and accounting periods, because it ignores any special transactions relating to the value of capital assets, e.g., property, aircraft. It must be remembered that by ignoring interest, taxes, Depreciation and Amortisation important charges against earnings have been removed from the calculation of earnings.

EBITDAR   these initials stand for Earnings Before Interest, Taxes, Depreciation, Amortisation, and Restructuring or Rent Costs, and it is an alternative measure of a company's financial performance. Its main value is a comparison between companies and accounting periods because it ignores any special transactions relating to the value of capital assets, e.g., property, aircraft and also any material changes to the way the business is run, as well as rents. It must be remembered that by ignoring all of these charges, significant charges against earnings have been removed from the calculation of earnings.

Equal instalment loan   a loan which is repaid by regular payment of the same amount that, in total, will fully repay the loan and the interest on it.

Equity   the money shareholders have invested in the company plus the profits which have not been paid out as dividends to shareholders. Equity is sometimes referred to as 'Shareholders' Funds'.

Estimated useful life   the estimated period, usually a number of years, during which an asset is expected to make a positive contribute to the company operations.

Executive Directors   a member of the Board of Directors who also has executive/management responsibilities within the company.

Exotic (derivative)   the name for a financial instrument which is more complicated than the simple and standard, plain vanilla, financial instrument.

Exposure draft   proposed version of a document, statement or standard which is issued prior to its release so that interested parties may comment before the final version is issued.

External Auditors   a firm of independent accountants employed to review and confirm the accuracy of a company's statutory accounts. In most countries an external auditor must be a member of an approved accounting organisation.

Finance Committee   a sub-committee of the Board of Directors responsible for overseeing and advising on the company's finances and financial risk management.

Finance lease   a lease contract where the owner of an asset permits a third party to use the assets in exchange for agreed periodic payments. During the period of the lease the third party (lessee) has all the benefits and obligations of ownership; this means the lessee must insure and maintain the assets as if they were their own. The finance lease document usually agrees how the asset will be dealt with at the end of the lease period. Often the assets will be leased to the lessee for the whole of its useful life.

Financial position   the status of a company's finances as shown in its financial report of assets and liabilities. Review of the detail should reveal possible areas for concern, e.g., a significant value of loans being repaid in a short period. The 'financial position' may also be referred to as a 'financial condition'.

Financial resources   there are different definitions, but the most common is, the total of a company's, liquid funds (cash), saleable securities, agreed arrangements by which the company can borrow funds in the future, also an assessment of the additional amount the company might be able to borrow, if necessary and prudent.

Financial strategy   The part of a strategy (or vision or target) which relates to the financial part of the airline's business.

Fixed budget   a detailed estimate for a period of expected revenue, expenses and resulting profit and capital transactions, which together give an estimated future financial result and financial position. The estimate is not changed at any time during the period.

Fixed-interest rate   an interest rate that does not change over the term or life of a loan, regardless of changes in the borrowing market.

Flexible budget   a detailed estimate for a period of expected revenue, expenses and resulting profit and capital transactions, which together give an estimated future financial result and financial position. The estimate for the period may be changed as the volume of actual transactions (e.g., passengers carried) becomes known.

Floating interest rate   an interest rate that moves up and down with the rest of the market or a defined index. In practice, the interest rate is re-fixed on stated dates, e.g., every three months.

Forecast   a prediction of future results based on the information which is currently known.

Forex   an abbreviation for 'Foreign Exchange', which is changing one national currency for another. Sometimes also written as 'FX'.

Forward sale   an agreement between a financial institution and a company by which the company agrees to sell an asset, in this circumstance an agreed amount of foreign currency, to the financial institution at a specified price and time in the future.

Fuel surcharge   an additional fee to cover an increase in fuel cost which is charged to passengers or shippers. Usually the additional charge has to be agreed by a government. Frequently it does not fully cover the additional extra fuel cost.

Full-Service Carrier (FSC)   an airline which offers more inflight and on the ground services than an LCC, and includes them in the airfare charged. Generally, an FSC operates to and from major airports, and does not separately charge for each service, although there are exceptions to this.

Many FSCs have adopted some aspects of the LCC's business model. An FSC will book passengers on other airlines and accept other airline's tickets. An FSC may also be called a 'Legacy Carrier'.

Fully costed   using an accounting method which calculates the total cost of providing a service or operating a route. May also be called 'absorption costing'.

Gearing   the ratio between a company's debt, i.e., the money borrowed from third parties such as banks, and its equity, i.e., the money shareholders have invested plus the profits which have not been paid out as dividends to shareholders. Equity is sometimes referred to as 'Shareholders' Funds'. The ratio may be called the 'debt/equity ratio' or the 'gearing ratio'.

Gearing ratio   the ratio between a company's debt, i.e., the money borrowed from third parties such as banks, and its equity, i.e., the money shareholders have invested plus the profits which have not been paid out as dividends to shareholders. Equity is sometimes referred to as 'Shareholders' Funds'. The ratio may be called the 'debt/equity ratio' or the 'gearing ratio'.

GHG   greenhouse gasses, i.e., a gas which is thought or proved to contribute to the greenhouse gas effect (i.e., trap radiated heat in the atmosphere).

Global Reporting Initiative an organisation that helps companies and government communicate their effect on non-financial matters (e.g., climate change).

Going Concern Basis a basis for preparing financial reports which assumes that a company has sufficient resources of money and staff to continue operating in the foreseeable future and to meet its obligation when they are due.

Greenhouse gas any gas which is thought or proved to contribute to the greenhouse gas effect, i.e., traps radiated heat in the atmosphere.

Independent Non-Executive Director a director of a company who does not have any managerial function or financial connection with the company other than receiving directors' fees. Sometimes referred to by the initials 'INED'. INEDs have the same responsibilities, duties and liabilities as other directors.

Information overload the situation where some much information is provided that it is difficult to understand the situation and establish whether the situation is or is not acceptable.

Interest cover a company's operating income before charging interest and tax (called EBIT), divided by the interest a company has to pay for a given period. The result is usually expressed as 'X times covered'. Example, EBIT = USD 10,000,000, interest payable = USD 2,500,000, calculation 10,000,000/2,500,000 = 4, expressed as '4 times covered'. When looking at the ratio as an indicator of risk, it is assumed that the lower the interest cover, the greater the possibility the company will not be able to pay its interest bill when due. Conversely, a higher interest cover is assumed to show that the company is more easily able to pay its interest bills when they are due.

Interim Results a set of statutory accounts produced for a period of less than 12 months, which are usually not audited, but have been subject to some form of independent check.

Internal controls systems and practices put in place within a company to safeguard assets and prevent fraud.

International Financial Reporting Standards (IFRS) accounting standards to be used for recording and reporting business transactions so that accounts are understandable and comparable across international borders. The standards are issued by the International Financial Reporting Board, an international standard-setting board.

International Integrated Reporting Council a council formed in 2010 with the objective of creating an international framework through which a company can explain its process for creating value over time.

Investment managers a person or company that acts on behalf of its clients, making investments of publicly quoted instruments, shares, bonds, etc., following the investment objectives and limits the client has agreed.

Investor Relations Department the department within a company responsible for dealing with investors, who may be shareholders or bond holders, with information on the company's financial performance and plans.

Jet Fuel Price Index   an index produced each week by IATA of the current price data for jet fuel from the information provider Platts.

Japanese Yen (JPY)   the legal currency of Japan.

Key Performance Indicator (KPI)   a set of measures a company uses to assess its progress towards meeting its strategic goals. The KPI will have both financial and operating measures. There are books written on how to relate KPIs to a company's strategy and how to use them in a company's management. The ratios may also be called Key Success Indicators or KSIs.

Lessee   a person or company that has agreed with the owner of an asset to use the asset in exchange for regular payments of money.

Lessor   a person or company which owns an asset who rents it to a lessee.

LIBOR   the interest rate at which banks offer to lend funds to each other in the international interbank market. The LIBOR is set every day by a number of international banks operating in London. They each state the rates at which they are prepared to lend to each other, and a calculation based on these rates produces the LIBOR.

Liquid funds   a collective term covering a company's cash holdings and securities which can be sold quickly and for which there is a generally quoted price.

Liquidation Basis   a basis for preparing financial reports which assumes the company will be, or is being, liquidated, i.e., is ceasing to trade and selling its assets and repaying its obligations.

Liquidity management   the methods used by a company to invest its liquid funds to make a return, but within the company's risk tolerance. The term can also be applied to the comparison of a company's liquid assets to its current liabilities to see whether there is a risk the creditors will not be paid on time.

Long-term plan   a plan which covers the longer-term future for operating a business, e.g., for an airline, number of passengers to be flown, when and where, maintenance to be done, etc., converted into money and presented to the board for approval and to be used as a general background for making decisions.

Low-Cost Carrier (LCC)   an airline whose approach is to minimising operating costs and relate as many costs as possible to the number of passengers carried. To achieve cost savings the airline may not provide some of the services offered by a Full-Service Airline, and may operate to and from less convenient airports. In addition, the airline may make a charge for some services, e.g., pre-boarding seat allocation, carry-on baggage. Generally, the airline will only issue tickets for its own flights. This business model may also be called a 'no frills' or 'budget airline'.

Management Accounts   the financial reports required by a company's directors and management to enable them to guide and manage the company's business.

Margin   the additional percentage added by a lender to a benchmark rate to determine the total interest rate the borrower must pay on a loan. The margin is frequently expressed a percentage (e.g., 0.75%) or as basis points (BPS), which are 1/100ths of a percentage (e.g., 75 BPS).

Money   the coins and notes which are legally approved to be used to settle debts and other transactions. Often now represented in electronic transactions.

Natural hedge   this is an approach to currency risk management where the receipts in a foreign currency are offset by expenses in the same currency, thus reducing the potential for loss due to changes in the foreign currency's exchange rate. If after offsetting the receipts by operating expenses in the same currency the surplus is larger than an airline's management wants, it may be possible to further reduce the surplus by arranging to finance some assets in the foreign currency.

Non-cash cost   a charge which is shown as an expense when calculating the profit, but is not actually paid in cash, e.g., Depreciation or an estimate for debtors who will not pay the amount due from them in full or in part.

Notes to the financial statements   comments attached to a company's statutory reports which give further detail of figures in the main reports or explain how the figures in the main reports have been calculated. The notes are designed to help the reader understand the content of the reports fully.

On-time performance   a measure of success for a transport service like an airline which has a published schedule. On-time performance is usually expressed as a percentage of its flights arriving or leaving on-time, or within a stated limit, compared to the total number of its flights arriving and departing. It is used as an indicator of efficiency.

Operating cash-flow   the amount of cash a company produces from the operation of its business.

Operating lease   a lease contract that allows a third party to use an asset but does not convey rights of ownership of the asset. Frequently the period for an operating lease is less than the asset's total useful life. Many airlines have part of their aircraft fleet on operating leases. Until the end of 2018 it was not a requirement to record operating leases in a company's balance sheet, but the obligations were detailed in a note to the balance sheet. The introduction of IFRS 16 requires all operating leases of more than one year to be recorded in a company's balance sheet as a liability and an accompanying asset. This change is predicted to be significant for airlines.

Operating profit   an accounting figure that calculates the profit for a period earned from all a company's business operations, excluding interest expenses for the period and taxes on profits. The figure is also referred to as EBIT (i.e., earnings before interest and tax).

Option contract   an agreement with a financial institution that grants a company the right, but not the obligation, to purchase or sell named assets, in this circumstance as an agreed amount of foreign currency, to the financial institution at an agreed price at an agreed time. Because the agreement does not impose an obligation to complete the transaction, the financial institution charges a premium to the company, which is paid whether or not the transaction is completed.

Overdraft   an agreement with a bank that permits a company to withdraw funds from its account with the bank even though the account does not have sufficient money to cover the amount withdrawn. When the bank account is in deficit it is said to be 'over-drawn'. There is always a limit to the amount an account can be over-drawn. This arrangement is a form of very short-term loan.

Payroll   the records need to calculate and arrange payment of the amounts due to the employees of a company.

Plain vanilla (finance)   the name for the simplest and standard version of a financial instrument (e.g., a bond or derivative).

Plan   a written set of agreed steps by which a future objective may be achieved.

Pledged   an asset held as security by a lender to guarantee payment of a debt or the completion of other forms of obligation. The security usually takes the form of a formal agreement between the owner and the lender agreeing that, in the event of the owner not repaying the loan, the lender will take over the asset. In addition, the agreement will be recorded in the ownership records for the asset, together with the lender's name and amount of the loan.

Premium (derivative)   a charge made by a company, usually a financial institution, when agreeing to grant an option to another company.

Price sensitive   any piece of information that, if made public, would be likely to have a significant effect on the price or value of a company's shares or bonds. In relation to a public company, if such information were to be released it should be to all investors so that all investors are treated equally.

Private placement   an arrangement and process through which a company raised funding by issuing share or bonds to an individual or small group of investors.

Profit   the amount remaining after the expenses required have been deducted from revenue earned from the operation of a business.

Profit and Loss Account   an account or statement which calculated the profit or loss produced by a company for a period. All of the company's revenue earned by operations is included together with all of the expenses related to generating the revenue.

Profitability   a class of financial ratios are used to assess a company's ability to generate earnings in excess of its expenses; the excess of earnings over expenses is called a 'profit'. A great deal has been written about profitability ratios. It may be appropriate to apply a separate set of profitability ratios to each separate business activity within a company, as well as to the company in total. The ratios are normally used to compare the company's and/or business activity's results with competitors or within a company when seeking to decide on the allocation of its resources.

Prospectus a document that describes a company, its operations, future prospects and risks that is given to prospective investors in a company's shares or bonds. If the shares or bonds are to be offered to the public, the details of the minimum information to be included in the prospects will be laid down by the stock exchange on which the company's share or bonds will be traded and listed.

Public Company a company whose shares are quoted on a Stock Exchange.

Public issue an arrangement and process through which a company raises funding by issuing shares or bonds to the public rather than to existing shareholders or bondholders. The Stock Exchange on which the shares or bonds will be quoted will usually require that the company issue a prospectus to potential investors, giving details of the issuer's business and the uses proposed for the money raised.

Public Relations Department the department within a company responsible for advising on action aimed at keeping good relationships between the company and the general public. This function may also be sub-contracted to a specialist company.

Qualified Opinion a statement issued by a company's external auditor, after they have examined the company's books of account and statutory reports, advising that the statutory reports have some material error. Generally, the opinion will state what, in the opinion of the auditor, the error or difference of opinion is.

Quick ratio a financial ratio designed to show whether a company is able easily to pay its current liabilities. The calculation is to add together all the current assets which can be converted into cash within a short time, perhaps 90 days, and divide this total by the total current liabilities. Calculating and using this ratio can be difficult, the company's balance sheet may not give sufficient information to assess how quickly current assets can be converted into cash; also within current liabilities are the loans which are payable within the following 12 months, but current liabilities do not include the cash expected to be generated in the next 12 months. Caution is needed when using this ratio. It is a more useful ratio for management.

Reconciliation of operating profit and operating cash-flow a statement which adjusted the operating profit shown in the Profit and Loss Account with the actual cash received from the operation of the company's business. It is necessary to make adjustments because the calculation of the operating profit includes some expenses which are not paid in cash (e.g., Depreciation), also, it assumes that all revenue has been received and all expenses have been paid, but there will be some unpaid revenue which will be in Debtors, and some expenses unpaid and these will be in Creditors.

Reporting Standards   national accounting standards to be used for recording and reporting business transactions so that accounts are understandable and comparable. The standards are usually issued by a national accounting standards board authorised by the government. The trend is for National Reporting Standards to mirror International Financial Reporting Standards.

Reserve Funds   an amount put aside to meet cash shortages in difficult times or in emergencies. A reserve fund is usually invested in highly liquid assets because the funds are likely to be required at short notice.

Residual value   an estimate of the amount of money an asset can sold for, either for scrap or to a third party, when it is no longer useful to the company.

Responsibility report   an approach used when presenting reports to a company's management. In the report each revenue and expense item is shown against the name or title of the manager who has direct control over the item, or a least understands what influences changes in the item if the company does not have direct control over it (e.g., rates of tax).

Return on Capital Employed (ROCE)   a financial ratio used to measure the earnings from the total funds a company has in use. The 'Capital' is a company's Shareholders' Funds plus its debt. The 'Return' is the company's profit from operations, excluding the interest and tax costs; this is the same as EBIT. The Return is expressed as a percentage of the Capital. The ROCE is used to assess whether the company's management is using its capital efficiently. The assessment is usually made by looking at the historic trend of the company's ROCE and how the recent ROCE compares with comparable companies. If a company's ROCE is deemed to be too low, the options are to increase the return without increasing the capital, or reduce the amount of capital without reducing the return, or a combination of the two. It is not unusual for the ROCE figures to be adjusted for unusual circumstances, for example, if a company has recently issued more shares but has not yet invested the funds raised.

Return on Equity (Return on Shareholders' Funds)   a financial ratio used to measure profit. It expresses the profit for the year as a percentage of the amount of money the company's shareholders have contributed to the company, either directly by paying money to the company or indirectly by permitting the company to keep part of the profits. The figure for Equity used in the calculation may be the year-end figure or the average for the year. Shareholders frequently compare the Return on Equity with the return that can be achieved by investing in other companies and other forms of investments (e.g., bonds).

Revenue Accounting   the records and process of recording the detail of the airline's actual revenue.

Revenue cost   the amount spent by a company on its day-to-day expenses (e.g., salaries, rent).

Risk an occurrence which may result in a monetary loss or bad result, e.g., damage to a person or company's reputation.

Rolling budget a detailed estimate for a period of expected revenue, expenses and resulting profit together with capital transaction, which together give an estimated future financial result and financial position. The estimate is only in detail for part of the period and the balance is in total. When the detailed period finishes, the total for the reminder of the period is revised in detail and an estimate for the beginning of the next period is added. In this way there is always a budget which rolls forward, the near-term in detail and further out in total.

Route contribution the amount contributed by an individual route to the total contribution. The calculation deducts the costs which can be solely identified to a route from the revenue generated by that route.

Route-operating cost the total operating cost of operating a particular route. The costs include those which can be identified solely to the route, and a share of all other operating costs apportioned to the route on some consistent basis.

SME these initials stand for Small and Medium Enterprise, which is a company that is in the middle (i.e., neither small nor large). When used in conversation, any definition for size can be applied (e.g., size of revenue, number of employees). When used in regulations or laws, a definition will be given. There is no internationally recognised definition.

Sale and lease back a transaction where an owner sells an asset and immediately leases it back from the new owner.

Self-insurance: a technique where a company accumulates funds to fund the cost of possible potential liabilities, instead of buying an insurance policy from a third party. This approach may also be called 'risk retention'.

Self-liquidating loan a loan with a term which is equal to or shorter than the period in which the borrower can complete a transaction or the useful life of an asset, and can use the proceeds of the transaction or the profit arising from the use of the asset to repay the loan.

Shareholder risk the risk that if the business operations of the company fail, partially or completely, shareholders will lose part or all of the funds they have invested in the company, both their initial investment and any profits which have not been paid to them.

Shareholders' Funds the money shareholders have invested in the company plus the profits which have not been paid out as dividends to shareholders. Shareholders' Funds are sometimes referred to as 'equity'.

Short-term forecast a revised estimate of the results of operating a business in the near-term. The period covered will depend on how far into the future acceptably accurate figures can be produced.

Short-term plan (budget) a plan which covers the near-term foreseeable future for operating the business, e.g., for an airline, number of passengers to be flown, when and where, maintenance to be done, etc., converted into money and presented to the board for approval so it can be used as a measure for the actual performance.

Solvent   the ability of a company to pay its debts as they become due; therefore a company must be confident it can repay its debts now and in the future. The inability to pay debts is called insolvency, and there may be legal consequences if a company continues to operate while insolvent.

Stakeholders   a person, group or organisation which is interested in or affected by the operation of a business.

Standby line of credit   an agreement with a bank or lending institution which permits a borrower to draw an agreed maximum amount of money from the bank or lending institution at short notice. In many respects it is like a normal loan, except it has pre-agreed terms.

Statement of cash-flows   a statement which summaries the actual cash-flows of a company for a period usually analysed into the cash-flow from,

- operating the business
- investing activities
- financing activities.

Statement of cash-flows   the statement included in a company's statutory accounts that, in total, shows all the company's inflows and outflows of cash split into 'Operations', 'Investing' and 'Financing'.

Statement of Changes in Equity   the statement included in a company's statutory accounts which shows the detail of changes during the accounting period in the amount which is due to the shareholders of the company. Typically, there are not many changes, often just the profit or loss for the year (profits increase the amount the shareholder invests and losses reduce it), and dividends paid to shareholders which reduces the amount the shareholders have in the company.

Statement of Principal Accounting Policies   the statement included in a company's statutory reports which describes the most important accounting policies it used to prepare the accounts for the period.

Statutory Accounting   the records required and the process of recording in order to produce the statutory reports.

Statutory Accounts   the financial reports a company is required to prepare and send to shareholders by the law of the country in which it is registered or operates, and/or the requirements of a regulatory organisation (e.g., a stock exchange). The accounts usually include a Profit and Loss Account, balance sheet (or Statement of Financial Position), and a cash-flow report and cover a 12-month period. The contents of the report are listed in the law or regulation, and may include non-financial information. The statutory accounts will be included with, and cross-referenced to, any other non-financial reports required by law or regulation.

Statutory reports   all the reports a company is required to produce, send to shareholders and lodge with government departments on its operations.

The reports include the statutory accounts. The contents of the reports are listed in the law or regulation and may include non-financial information. The statutory accounts will be included with, and cross-referenced to, any other non-financial reports required by law or regulation.

Stock exchange   a regulated market where company shares and other forms of investments can be bought and sold at prices agreed by the seller and buyer.

Stocks   a quantity of goods held for future use.

Strategy   a high-level statement which describes and defines the long-term objectives of the airline approved by the airline's Board of Directors. It may also be called a 'Vision' or 'Target'.

Success measures   a set of measures a company uses to assess its progress towards meeting its strategic goals. The 'success measures' will have both financial and operating measures. See also Key Performance Indicators.

Sustainability Accounting Standard Board   an organisation formed to design Reporting Standards related to social and environmental matters, and to identify the information that is most useful when making decisions.

Syndicate   a group of individuals or companies who co-operate and combine resources in order to make a large loan to a company.

Syndicated loan   a loan made by a group or syndicate of lenders to a single borrower, with one of the group authorised to negotiate on behalf of the whole group.

Tankering   uplifting onto an aircraft more jet fuel than is needed for a flight because the cost of the fuel is significantly cheaper than it is at other airports. The fuel is then used on subsequent flights. The calculation of whether it is worthwhile to 'tanker' are complicated because of the extra cost of carrying the extra fuel.

Target   a high-level statement which describes and defines the long-term objectives of the airline approved by the airline's Board of Directors. It may also be called a 'Strategy' or 'Vision'.

Taxation   a charge made by a government or authority authorised by a government.

Times Assets Turned Over   a financial ratio used to indicate how efficiently a company's assets are being used. The calculation is to divide the company's total revenue earned by the total value of assets in the business (i.e., Revenue/Total Assets = times). In some calculations, the value of total assets is the average of the total assets at the beginning of the year and the total assets at the end of the year. Generally, the higher the number of times the more efficient the company is thought to be. The ratio is mostly used to make comparisons with other companies in the same industry with similar business models.

Trading limits   a limit set by a company on the total value of transactions that may be outstanding with another company at any one time.

Treasury   the function of managing the company's cash-flow, arranging borrowings, the investment of short- and long-term funds in deposits and investments. Also managing financial risks.

True and Fair View   means that, after examining the statutory accounts, the external auditors think the financial statements are substantially correct, i.e., free from material misstatements and faithfully represent the financial performance and position of the company.

Unearned Transportation Revenue   the account recording the amounts received in advance from passengers and shippers for flights and/or cargo shipments. When the passenger travels or the cargo is carried, the appropriate amount will be transferred from this account to the airline's 'Profit and Loss Account' as revenue. If a passenger fails to travel or complete the full itinerary, they may or may not be entitled to a refund. If no refund is due, the amount will be treated as revenue.

USD: United States Dollar   the legal currency of the United States of America.

Value at Risk   a technique which seeks to measure the value of risk in a company's operations.

Variance   the difference between the actual result and the budget for the same period.

Venture capital   investment provided by firms or individuals to newly formed companies which are in their early stages of development and operation. The funds are provided to the newly formed company by buying newly issued shares.

Venture capitalist(s)   companies, individuals, or groups of individuals who invest in newly formed companies, taking the significant initial risk of loss if the new company fails or does not meet its initial targets. Their expectation is to be able to sell their shares at an early stage, achieving a high rate of return on the investment.

Vision   a high-level statement which describes and defines the long-term objectives of the airline approved by the airline's Board of Directors. It may also be called a 'Strategy' or 'Target'.

Wear and tear   the deterioration or change in condition of an asset which is to be expected following its normal use.

'Wet' lease   an agreement by which the owner of an asset, e.g., an aircraft, agrees to lease the asset to a third party, together with the personnel required to operate the asset. In the case of an aircraft, this would be the cockpit crew and may or may not include cabin crew.

Windfall profit   an unexpected profit, usually arising from some unexpected source or unforeseen circumstances.

Working capital   a general term describing the funds used operating a company's day to day business. In the company's balance sheet, 'Working Capital' is the difference between the company's Current Assets, e.g., Stocks, Debtors and its liabilities (e.g., creditors). In many loan agreements there is a requirement that the borrower should ensure it always has 'positive working capital' (i.e., that its Current Assets exceed its Current Liabilities).

Index

terms used in the financial strategy 16–19
transfer of value 49
types of reports *see* reports

users of reports 89–90

variance analysis 87

working capital management 57–61;
measured and monitored 58;
responsibility 58

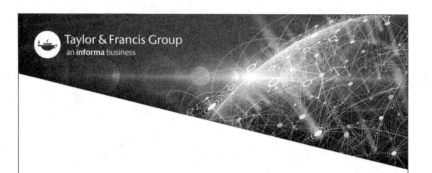

Printed in the United States
by Baker & Taylor Publisher Services